337

Topics in Current Chemistry

For further volumes:
http://www.springer.com/series/128

Aims and Scope

The series Topics in Current Chemistry presents critical reviews of the present and future trends in modern chemical research. The scope of coverage includes all areas of chemical science including the interfaces with related disciplines such as biology, medicine and materials science.

The goal of each thematic volume is to give the non-specialist reader, whether at the university or in industry, a comprehensive overview of an area where new insights are emerging that are of interest to larger scientific audience.

Thus each review within the volume critically surveys one aspect of that topic and places it within the context of the volume as a whole. The most significant developments of the last 5 to 10 years should be presented. A description of the laboratory procedures involved is often useful to the reader. The coverage should not be exhaustive in data, but should rather be conceptual, concentrating on the methodological thinking that will allow the non-specialist reader to understand the information presented.

Discussion of possible future research directions in the area is welcome.

Review articles for the individual volumes are invited by the volume editors.

Readership: research chemists at universities or in industry, graduate students.

Judith Klinman • Sharon Hammes-Schiffer
Editors

Dynamics in Enzyme Catalysis

With contributions by

K. Arora · C.L. Brooks · C.M. Cheatum · Q. Cui · M.D. Daily · K.W. Harpole · V.J. Hilser · D. Hilvert · A. Kohen · B. Ma · V.R. Moorman · R. Nussinov · G.N. Phillips · T.P. Schrank · C. Schulenburg · S.D. Schwartz · A.J. Wand · J.O. Wrabl · H. Yu

Springer

Editors
Judith Klinman
University of California
Department of Chemistry
Department of Molecular and
Cell Biology, and the California
Institute for Quantitative Biosciences
Berkeley, CA
USA

Sharon Hammes-Schiffer
Department of Chemistry
University of Illinois at
Urbana-Champaign
Urbana, IL
USA

ISSN 0340-1022 ISSN 1436-5049 (electronic)
ISBN 978-3-642-38961-0 ISBN 978-3-642-38962-7 (eBook)
DOI 10.1007/978-3-642-38962-7
Springer Heidelberg New York Dordrecht London

Library of Congress Control Number: 2013944058

Printed on acid-free paper

Springer is part of Springer Science+Business Media (www.springer.com)

Foreword

The enormous number of three-dimensional structures for proteins that emerged during the second half of the twentieth century – as of November, 2012, there are ca. 86,000 entries in the international Protein Data Base – led quite naturally to an intense focus on the relationship between protein structure and function. There was recognition that allosteric proteins could assume different shapes in the presence of allosteric regulators, and much work was dedicated to developing mathematical models for this behavior [1,2]. The classical studies on hemoglobin showed how the remote binding of small molecules could tune the affinity of molecular oxygen at the active site heme-iron center [3]. These protein motions were conceptualized as rigid body movements that linked the active and inactive protein states. The latter half of the twentieth century saw an increasing focus on the more general feature of protein dynamics. Once again, the binding of small gaseous molecules to heme proteins played an important role, with Frauenfelder and his co-workers demonstrating a photo-initiated, multi-exponential release of CO from myoglobin at low temperature [4]. The implication from these studies was that the inherent flexibility of proteins at room temperature had led to a family of low-temperature conformers with different inherent kinetic properties. Concomitant with experimental approaches, Karplus and co-workers developed methods for computer simulations of rapid protein motions (picosecond timescale) using Newton's equations of motion [5]. For many years, classical studies of enzyme function and the study of protein dynamics moved on largely parallel trajectories, occasionally intersecting but with considerable skepticism and reservation on both sides.

The last decade has seen a profound shift toward the recognition that the large size and inherent floppiness of proteins can manifest itself at every level of protein function. This type of motion goes beyond two state models for allostery, as well as the expected accommodation of a protein's shape to the binding of substrates and their conversion to products. The emerging models include a productive and causative role for protein motions at every level of function. This volume highlights some of the exciting directions that the field of protein dynamics has made in recent years. Here, the term "dynamics" refers to both equilibrium and non-equilibrium motions on all timescales. We have aimed at a balanced presentation of four chapters on theory and an equal number on experimental approaches. This is a very active research area, and we apologize upfront to the many investigators who are not included in the present volume. Our hope was that the range of chapters chosen would highlight topics that may not always be considered in the same context.

From the experimental perspective, the role of protein dynamics in enzymatic bond-making and bond-cleavage events is addressed by Kohen and Cheatum. As has been discussed with increasing frequency in the literature, the enzymatic activation of C–H bonds opens up a special window for analyzing the role of heavy atom protein motions in catalysis [6]. Schulenberg and Hilvert address the very active area of intrinsically disordered proteins, and whether the properties of such enzymes can help us to understand the role of protein motions in native enzyme catalysis. The correlation of side chain entropy among protein side chains with function has increasingly attracted wide attention, and Wand and co-workers describe careful and insightful studies of this feature of calmodulin function according to an "entropy ruler." Finally, Schrank, Wrabl, and Hilser talk about the role of a lid closure on the function of adenylate kinase, building upon multi-layered experimental data to generate a multi-state model for the progression of this enzyme from an open to a closed, catalytically active state.

From the theoretical perspective, the effects of macromolecular crowding on enzyme conformational dynamics and activity are discussed by Ma and Nussinov. Structured crowding, which refers to the highly organized cellular environment, could potentially enhance enzyme efficiency and specificity. Cui and coworkers discuss how atomistic and coarse-grained simulations can be used to elucidate the mechanism of activation transitions, which often precede the chemical steps in enzymes and are typically allosteric and multiscale in nature. In the simplest enzymes, such as adenylate kinase, the activation transition corresponds to the closure of the active site, but in biomolecular motors, such as myosin, it is more complex. Arora and Brooks focus on the enzyme dihydrofolate reductase (DHFR), which catalyzes the hydride transfer reaction required for the conversion of dihydrofolate to tetrahydrofolate. This enzyme has been studied extensively with a wide range of theoretical and experimental methods. The consensus is that equilibrium conformational motions of the protein and ligands impact catalysis in DHFR by generating configurations conducive to the hydride transfer reaction, but there is no evidence for dynamical coupling of protein vibrational modes to the breaking and forming of chemical bonds [7]. Schwartz proposes a more controversial perspective about the role of protein dynamics in enzyme catalysis. While he agrees with the above interpretation for DHFR, he also proposes that femtosecond to picosecond "promoting vibrations" of the protein can be directly coupled to the passage over the chemical barrier (i.e., to bond breaking/forming) and are catalytically relevant for other enzymes, such as lactate dehydrogenase and purine nucleoside phosphorylase. This aspect of catalysis contrasts with the more generally accepted model of equilibrium conformational sampling that has components of protein motions ranging from femtosecond to millisecond timescales.

As the reader will see, the included chapters are, at times, either complementary or divergent in their conclusions. We believe that this reflects the current vibrancy of the field of protein dynamics. One of the goals of this community – hopefully achievable in the coming decades – will be to develop a level of understanding that allows us to replicate in the test tube the exquisite specificity and enormous catalytic rate accelerations of native proteins.

References

1. Monod J, Wyman J, Changeux JP (1965) On nature of allosteric transitions – a plausible model. J Mol Biol 12:88–188
2. Koshland DE, Nemethy G, Filmer D (1966) Comparison of experimental binding data and theoretical models in proteins containing subunits. Biochemistry 5:365–368
3. Edelstein SJ (1975) Cooperative interactions of hemoglobin. Annu Rev Biochem 44:209–232
4. Hong MK, Braunstein D, Cowen BR, Frauenfelder H, Iben IET, Mourant JR, Ormos P, Scholl R, Schulte A, Steinbach PJ, Xie AH, Young RD (1990) Conformational substates and motions in myoglobin – external influences on structure and dynamics. Biophys J 58:429–436
5. Brooks BR, Bruccoleri RE, Olafson BD, States DJ, Swaminathan S, Karplus M (1983) Charmm – a program for macromolecular energy, minimization, and dynamics calculations. J Comput Chem 4:187–217
6. Klinman JP (2009) An integrated model for enzyme catalysis emerges from studies of hydrogen tunneling. Chem Phys Lett Fontiers 471:179–193
7. Hammes-Schiffer S, Benkovic SJ (2006) Relating protein motion to catalysis. Annu Rev Biochem 75:519–541

Contents

Top Curr Chem (2013) 337: 1–40
DOI: 10.1007/128_2012_407

Published online: 29 March 2013

Relationship of Femtosecond–Picosecond Dynamics to Enzyme-Catalyzed H-Transfer

Christopher M. Cheatum and Amnon Kohen

Abstract At physiological temperatures, enzymes exhibit a broad spectrum of conformations, which interchange via thermally activated dynamics. These conformations are sampled differently in different complexes of the protein and its ligands, and the dynamics of exchange between these conformers depends on the mass of the group that is moving and the length scale of the motion, as well as restrictions imposed by the globular fold of the enzymatic complex. Many of these motions have been examined and their role in the enzyme function illuminated, yet most experimental tools applied so far have identified dynamics at time scales of seconds to nanoseconds, which are much slower than the time scale for H-transfer between two heavy atoms. This chemical conversion and other processes involving cleavage of covalent bonds occur on picosecond to femtosecond time scales, where slower processes mask both the kinetics and dynamics. Here we present a combination of kinetic and spectroscopic methods that may enable closer examination of the relationship between enzymatic C–H → C transfer and the dynamics of the active site environment at the chemically relevant time scale. These methods include kinetic isotope effects and their temperature dependence, which are used to study the kinetic nature of the H-transfer, and 2D IR spectroscopy, which is used to study the dynamics of transition-state- and ground-state-analog complexes. The combination of these tools is likely to provide a new approach to examine the protein dynamics that directly influence the chemical conversion catalyzed by enzymes.

Keywords Enzyme Dynamics · Hydrogen Tunneling · 2D IR · Vibrational Spectroscopy · Isotope Effects

C.M. Cheatum and A. Kohen (✉)
Department of Chemistry, University of Iowa, Iowa City, IA 52242, USA
e-mail: amnon-kohen@uiowa.edu

Contents

Abbreviations

2D IR	Two-dimensional infrared spectroscopy
CLS	Centerline slope
FFCF	Frequency–frequency correlation function
KIE	Kinetic isotope effects
$PAAD^+$	Picolyl azide adenine dinucleotide

1 Introduction

1.1 General

First, we wish to state up front that, like most researchers studying enzyme-catalyzed reactions, we do not address "catalysis," i.e., the ratio of the rates of the catalyzed reaction and the uncatalyzed reaction. The reason is mostly practical, namely that enzyme catalyzed C–H activations do not commonly have a relevant uncatalyzed model reaction with which to compare. A few exceptions where model reactions were examined are presented in [1] and references cited therein. Below, we address studies of the enzyme-catalyzed reaction, and, more specifically, redox reactions involving hydride transfer.

The focus of this review is on the role of fast motions – on the femtosecond to picosecond time scale – in the H-transfer involved in C–H activation by enzymes catalyzing hydride, hydrogen, and proton transfer. The role of large motions of the protein, on the time scale of enzyme turnover (i.e., milliseconds to seconds), is well established and broadly accepted by the scientific community. These motions are mostly associated with substrate binding (i.e., "induced fit") and product release. Studies of protein dynamics on faster time scales (microsecond to femtoseconds), on the other hand, are still in their infancy, and no single working model is broadly accepted by all researchers. While there is little doubt that proteins, like any other molecule at physiological temperatures, move and fluctuate on these fast time scales, the role of these motions in the catalyzed reaction is not fully understood. Different studies, both experimental and theoretical, lead to diverse conclusions, commonly based on the focus, strengths, and limitations of the methods applied. For example, NMR relaxation experiments examine the environmental dynamics that are coupled to the spins under study at microsecond to nanosecond time scales (μs–ns). These measurements are lengthy and use the apo-enzyme or a stable enzymatic complex that may or may not resemble one of the kinetic intermediates along the catalytic path. These studies of protein dynamics naturally monitor motions that are in thermal equilibrium ("statistical dynamics") [2] and exclude examination of motions that may not reach thermal equilibrium at the transition state. The same limitation is common to computer simulations focused on the potential of mean force (PMF) along the reaction coordinate. Nevertheless, recent calculations suggest that the relaxation to thermal equilibrium in a protein approaching the transition state may require 100 fs or more [2]. Since some chemical events are faster than that time scale (e.g., QM tunneling), it is not possible to exclude the possibility that nonequilibrium fluctuations of the enzyme as it approaches the transition state may affect the chemistry. It is difficult to examine experimentally dynamics that are not in equilibrium (i.e., "non-statistical" or "nonequilibrium" dynamics) let alone determine whether such dynamics play a role in the catalyzed reaction. 2D IR spectroscopy measures the relaxation of equilibrium fluctuations in the enzyme active site which, within the fluctuation–dissipation theorem, should also reflect the time scale for relaxation of nonequilibrium fluctuations. Concluding that nonequilibrium motions can persist on time scales longer than the chemical step, however, is neither evidence that such motions exist nor that they play any role in controlling the chemical kinetics.

Most kinetic models and rate theories, including those used by us, if they are sufficiently complex, can rationalize, and sometimes even qualitatively fit, the experimental kinetics, without invoking "nonequilibrium dynamics." Frequently these models and mathematical expressions are complex and involve more parameters than experimental data points. For non-adiabatic systems (most hydrogen radical transfer, or proton coupled electron transfer [3]), the C–H bond is not activated at all until the system is rearranged to the tunneling ready state (TRS, see more below). Although the rearrangement of the system toward TRS might be on the microsecond time scale, the actual bond activation occurs on the femtosecond time scale. Consequently, methods to study either statistical or nonequilibrium dynamics that are affecting, sensed by, or coupled to the reaction must be able to probe those motions on the chemically relevant time scale, femtosecond to picoseconds. Below we present studies attempting to measure such dynamics and correlate these to the nature of the H-transfer step and its kinetics. Since the dynamics discussed below are measured spectroscopically, they require examination of a stable state. We try to examine states involving a TS-analog or other enzymatic complexes relevant to the reaction under study. Then we will try to pursue correlation between these fast dynamics, and kinetic properties, or the nature of the fast H-transfer step in the same enzymatic system.

1.2 Kinetics

The nature of the C–H → C transfer and its TS can be probed by three experimental approaches: measurements of the reaction rate (related to $\Delta G^{\ddagger}$), kinetic isotope effects (KIEs, related to $\Delta\Delta G^{\ddagger}{}_{i-j}$, where i and j are the light and heavy isotopes, respectively), and linear free energy relationships (relating $\Delta G^{\ddagger}$ to ΔG°, which will not be discussed further in this review). This review will focus on KIEs as a means to assess the nature of the enzyme catalyzed H-transfer. An advantage of measuring KIEs on the chemical step, as opposed to rate measurements, is that they connect more easily to calculations and simulations. Most theoretical simulations of enzyme catalyzed reactions focus only on a single step, i.e., the TS of the bond cleavage and the two stable states that it connects (reactant and product states). The absolute $\Delta G^{\ddagger}$ calculations suffer from the fact that the contributions to the free energy involve hundreds of kcal/mol, but most enzymatic $\Delta G^{\ddagger}$s are below 20 kcal/mol. Thus, the accuracy of the unparameterized rate calculation is often quite low. KIEs, on the other hand, are only sensitive to the few isotopically sensitive parameters and often involve the fortuitous cancelation of many of the inherent errors in the calculated free energies. Thus, KIEs provide a much more reliable assessment of the accuracy of the calculated PMF and the predicted progress of the reaction over and through the barrier.

From a kinetic point of view, examination of the chemical step (i.e., covalent bond cleavage and formation) in the complex kinetic cascade of most enzymes is a major challenge. Since most enzymes evolved under conditions other than saturation of their substrate, the turnover number measured for enzymes (k_{cat}, measured under substrate saturation) is frequently rate limited by product release rather than by the chemistry. Under steady state conditions, the second-order rate constant in enzymatic reactions (k_{cat}/K_M, measured at very low substrate concentration) is also not typically rate limited by the chemical step, but rather by events associated with substrate binding, or induced fit motions (e.g., large scale conformational changes that follow the binding step), etc.

From an experimental point of view KIEs are measured using two basic approaches: (1) non-competitive KIEs and (2) competitive KIEs. The first method consists of rate measurements with two different isotopologs (i.e., reactants that only differ in their isotopic composition), followed by calculation of the ratio of the rate constants. The second method directly measures the KIE through the competition of both isotopologs in the same vessel by following the enrichment of the heavier isotope in the reactants or its depletion in the product. Although the second approach doesn't provide rate information and is limited to measurements of the KIE on the second order rate constant [4], it is also not subject to artifacts involving different contaminations or different conditions in the two measured rates that are common problems in the first method. Additionally, the competitive approach affords the use of trace amounts of tritium (competing with H or D), eliminating the need for "carrier free" tritiated reactant (i.e., pure T with no H or D contamination) that would be needed for non-competitive tritium KIEs. This last issue indeed dictates that when triple isotopic labeling is required for the application of Northrop's method [see Eq. (2)], a competitive KIE measurement is commonly used.

1.2.1 Kinetic Complexity

In studying C–H → C transfer, the advantage of KIEs is that the isotopic label is placed on a C–H bond that has little contribution to binding events but has a substantial effect on the H-transfer, the potential energy surface along the reaction coordinate, the chemical character of the TS, and the nature of the barrier crossing which can all be examined. Such KIEs involve hydrogens that are transferred in the chemical step (i.e., 1° KIEs) or are changing bond order during the transfer of another hydrogen (i.e., 2° KIEs). Unfortunately, observed KIEs are often "masked" by kinetic complexity. An example of the relationship between the observed KIE and its intrinsic value (assuming only the chemical step is isotopically sensitive) is given below for the second order rate constant, which is used in most of the case studies presented below.

When measuring KIEs, one has to bear in mind that the observed KIEs can be obscured by "kinetic complexity," where partially rate-limiting steps that are not isotopically sensitive diminish the observed KIE relative to the value of the "intrinsic" KIE on the step of interest [5]. The observed KIE ${}^{i}(V/K)_{j}$ where j is the lighter hydrogen isotope and i is the heavier one) is related to the intrinsic KIE by the equation

$$ {}^{i}(V/K)_{j} = \frac{(k_j/k_i)_{\text{int}} + C_{\text{f}} + C_{\text{r}} \cdot \text{EIE}}{1 + C_{\text{f}} + C_{\text{r}}}, \qquad (1) $$

where $(k_j/k_i)_{\text{int}}$ is the intrinsic KIE and EIE is the equilibrium isotope effect or the isotope effect on the equilibrium constant of the reaction. The terms C_{f} and C_{r} are the forward and reverse kinetic commitments to catalysis, respectively. C_{f} reflects the ratio of the forward rate of the isotopically sensitive step (the step of interest) to the net reverse rate of the non-isotopically sensitive steps that precede this step [6].

Similarly, C_r reflects the reverse rate of the isotopically sensitive step to the net forward rate of the non-isotopically sensitive steps that follow this step. For an irreversible reaction, $C_r = 0$ and all the "kinetic complexity" is caused by C_f from the preceding steps. Analyzing the kinetic commitments can sometimes provide information on the reaction mechanism, as will be seen in the context of TSase below.

Experimentally, examination of the chemical step is challenging because most mature wild-type enzymes, when catalyzing the reaction with their natural substrate under physiological conditions, are not rate limited by the chemical step. With careful experimental design and/or isotopic labeling, the chemical step is the only isotopically sensitive one, as in the comparison of C–H to C–D bonds cleaved in large biological molecules that are not otherwise different. In many cases, exposure of the isotopically sensitive step also requires the use of non-natural substrates, an active site mutation(s), or both (e.g., alcohol dehydrogenases) [7].

Pre-steady state measurements are another way to remove some of the kinetic complexity. These experiments are designed to exclude effects from substrate binding and product release. These measurements focus on kinetic steps between the formation of the complex where the enzyme is already bound to all reactants (typically a binary or ternary complex), and the formation of the enzyme-bound product. Nevertheless, the measured rates often reflect the impact of other effects including conformational changes, pK_a and protonation shifts, and other microscopic kinetic contributions that can still mask the chemical step. For example, many nicotinamide-dependent enzymes bind the adenine–pyrophosphate moiety of this cofactor first, followed by major conformational changes of the protein and rotation of the reactive nicotinamide ring into the active site prior to reaching the ground state of the chemical step.

For systems where these solutions are not effective, an instrumental approach is to use Northrop's method for finding intrinsic KIEs based on a combination of deuterium and tritium KIEs as is represented by the equation [8, 9]

$$\frac{{}^{\mathrm{T}}(V/K)_{\mathrm{D}}^{-1} - 1}{{}^{\mathrm{T}}(V/K)_{\mathrm{H}}^{-1} - 1} = \frac{(k_{\mathrm{H}}/k_{\mathrm{T}})_{\mathrm{int}}^{-0.2994} - 1}{(k_{\mathrm{H}}/k_{\mathrm{T}})_{\mathrm{int}}^{-1} - 1}. \tag{2}$$

In this equation ${}^{\mathrm{T}}(V/K)_{\mathrm{D}}$ and ${}^{\mathrm{T}}(V/K)_{\mathrm{H}}$ are the observed D/T and H/T isotope effects on the second order rate constant k_{cat}/K_{M}, respectively, and $(k_{H}/k_{T})_{int}$ is the intrinsic H/T KIE on the chemical step of interest. Although this method uses the Swain–Schaad relationship and relies on several assumptions [10], its applicability to systems where tunneling is known to be important has been addressed and confirmed experimentally [9]. A simple way to understand the Northrop method is that the reference isotope [tritium in Eq. (2)] affords the intrinsic comparison of the ratio of k_H to k_D, i.e., the intrinsic KIE. It is important to note that the reference isotope is not limited to tritium [as in Eq. (2)] but any of the three isotopes of hydrogen could serve as the reference [9].

1.2.2 Temperature Dependence of KIEs

The temperature dependence of KIEs (TDKIEs) reveals the differences in the enthalpies and entropies of activation for the two isotopes ($\Delta\Delta H^{\circ}$, or ΔE_{a} and A_j/A_i, respectively). Since both isotopes share the same potential energy surface, these activation parameters reflect the nature of the reaction coordinate under study. This temperature dependence has been analyzed by a variety of theoretical approaches. Semiclassically, ΔE_{a} reflects the difference in zero-point-energies (ZPE) for the two isotopes at the ground state (GS) as compared to the transition state (TS), and A_j/A_i should be close to unity [11, 12]. If a tunneling correction is added to the semiclassical model, it can be calculated using either a simple parabolic approximation (e.g., the Bell correction) [13] or assuming more complex potential surfaces [14]. Although such corrections can explain many observations, they cannot address other features such as temperature independent KIEs with a significant energy of activation. One model that enables a simple fit of such data is the Limbach–Bell correction, in which the Bell tunneling correction to the TST rate [13] is further corrected by fitting the mass of the transferred particle as a parameter [15]. Fitting this model to temperature independent KIEs with significant ΔE_{a} results in isotopic masses that are much higher than those of H, D, or T. These large masses are in accordance with coupling of the transferred particle to larger system, but provide limited physical meaning. Approaches that separate the temperature dependence of the rate from that of the KIEs reflect more realistic, multidimensional models of the reaction and are all summarized here as the Marcus-like model.

Marcus-like models assume a priori that the hydrogen atom is light enough that a full quantum mechanical treatment including quantum tunneling is necessary to describe the reaction adequately. The prevailing model of reactivity, then, is an extension of the Marcus theory of electron tunneling [16], where the tunneling particle is a hydrogen rather than an electron. This model, which has been described under different names (Marcus-like models [17], environmentally coupled tunneling [18], full-tunneling model [19], vibrationally enhanced tunneling [20], etc.), proposes a mechanism where heavy atom reorganization leads to a "Tunneling Ready State" (TRS) in which the quantum states of the hydrogen in the reactant and product wells are degenerate and, thus, the hydrogen can transfer to products via tunneling. This TRS becomes the transition state for H-transfer. Several groups have independently derived a functional form for the rate constant (k) for an H-transfer reaction within this model, which is given by the expression [3, 18, 21–23]

$$k = C(T)\frac{|V|^2}{\hbar}\sqrt{\frac{\pi}{\lambda k_{\mathrm{B}}T}}\mathrm{e}^{-\frac{(\Delta G^{\circ}+\lambda)^2}{4\lambda k_{\mathrm{B}}T}}\int_0^{\infty} F(m,\mathrm{DAD})\,\mathrm{e}^{-\frac{E(\mathrm{DAD})}{k_{\mathrm{B}}T}}\,\mathrm{dDAD}. \qquad (3)$$

In this equation the terms in front of the integral give the standard Marcus theory [16] rate for reaching the TRS (Fig. 1a) based on the electronic coupling (V), the reorganization energy (λ), the driving force for the reaction (ΔG°), and the absolute temperature (T). The additional term $C(T)$ is the fraction of reactive complexes [24]

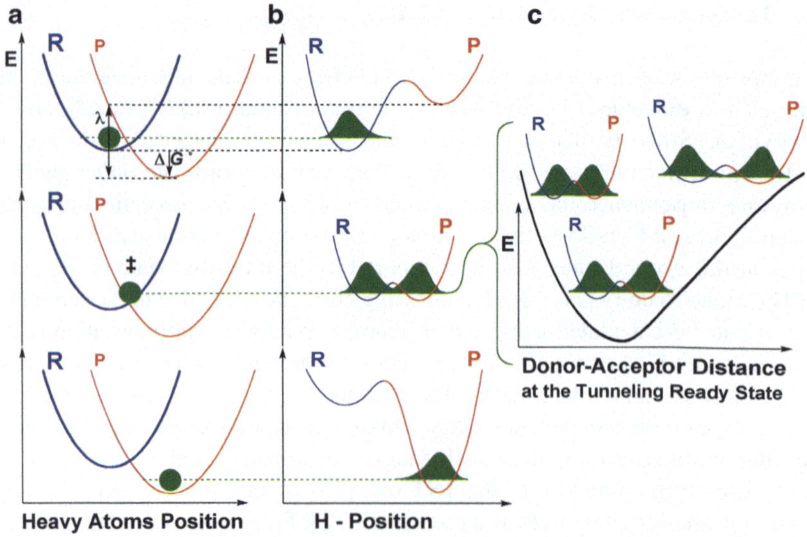

Fig. 1 Marcus-like models of hydrogen tunneling. Three slices of the potential energy surface (PES) along components of the collective reaction coordinate showing the effect of heavy-atom motions on the zero point energy in the reactant (*blue*) and product (*red*) potential wells. (**a**) The heavy atoms coordinate. (**b**) The H-atom position. In the *top panel*, the hydrogen is localized in the reactant well, and the energy of the product state is higher than that of the reactant state. Heavy atom reorganization brings the system to the tunneling ready state (TRS, middle state in (**a**, **b**)), where the zero point energy levels in the reactant and product wells are degenerate and the hydrogen can tunnel between the wells, for short donor–acceptor distances (DADs). (**c**) The effective potential surface (EPS) along the DAD coordinate at the TRS. Further heavy atom reorganization breaks the transient degeneracy and traps the hydrogen the product state (*bottom panel*). The rate of reaching the TRS depends on the reorganization energy (λ) and driving force (ΔG°), which are indicated in the *top panel*, and further discussed under Eq. (3). The effects of DAD sampling on the wavefunction overlap in the TRS. Tunneling probability is proportional to the overlap integral of the hydrogen wavefunctions in the reactant (*blue*) and product (*red*) wells, which depends on the DAD

to account for the conformational landscape that brings the H-donor and acceptor to their chemical ground state. The terms in front of the integral are essentially insensitive to the mass of the transferring particle, so the KIE is determined by factors within the integral. The integral gives the probability of tunneling from the reactant well to the product well once the system reaches the TRS (the middle state in Fig. 1a, b). The first term in the integral is the probability of tunneling for a particle with mass *m* as a function of the donor–acceptor distance (DAD). This probability is multiplied by a Boltzmann factor (second term) that gives the probability of finding the system at a particular DAD. This product is integrated over all DADs to yield the total tunneling probability. An important characteristic of this model is that the entire system is assumed to be in thermal equilibrium. In many previous publications the use of the word "dynamics" simply referred to "motions" without distinguishing between motions that are in thermal equilibrium (i.e., "statistical dynamics") and those that are not, i.e., "non-statistical dynamics" [2, 17, 19, 25, 26]. This imprecise language subsequently led some to believe that a controversy exists over whether true non-equilibrium dynamics contribute to enzyme-catalyzed reactions [27, 28]. Although several such proposals do exist [2], the Marcus-like models do not require and thus do not invoke non-equilibrium dynamics.

1.3 2D IR Spectroscopy

Assessment of the potential for a functional role for fast motions of the protein in the catalyzed reaction requires a way to characterize enzyme motions and to identify specifically those motions that directly affect the reactive complex when it is at the TRS. Two-dimensional infrared spectroscopy (2D IR) is a spectroscopic tool that has emerged over the past decade and has seen applications to a wide range of problems [29]. 2D IR can directly characterize protein dynamics at the enzyme active site at the time scale of hundreds of femtoseconds (fs) to tens of picoseconds (ps). Thus, 2D IR spectroscopy allows direct observation of the protein dynamics at the enzyme active site at the time scale that is most relevant for the catalyzed C–H → C transfer.

2D IR spectroscopy is an optical analog of 2D NMR spectroscopy. Specifically, when applied to problems involving dynamics, 2D IR measures correlations in a way that is directly analogous to a COSY experiment. The 2D IR spectrum is two-dimensional in that it displays the intensity of the signal as a function of two frequencies: the excitation frequency, ω_1, and the detection frequency, ω_3. Figure 2 shows an example of a 2D IR spectrum of azide in D_2O. There are three types of signal contributions giving rise to two oppositely signed peaks in the 2D IR spectrum. Both of these peaks result from the change in absorbance of the sample as a result of the interactions with the excitation pulses. The excitation moves population from the ground state of the oscillators to the vibrationally excited state, and the probe pulse measures the change in absorbance resulting from the excitation. In the ω_1 axis, both peaks appear at the fundamental transition frequency.

Moving population out of the ground state into the first excited state causes a decrease in the absorbance of the probe beam. This signal is often referred to as a ground-state-bleach signal. In addition, the excited population can undergo stimulated emission as a result of the interaction with the probe pulse, giving rise to a signal that also behaves like a decrease in absorbance. Together, the ground-state bleach and stimulated-emission signals produce the red peak centered on the diagonal ($\omega_3 = \omega_1$), which, when plotted in transmittance, is a positive peak (decreased absorbance corresponds to increased transmittance). The other peak in the 2D IR spectrum (blue) results from the excited-state-absorption of the vibrationally excited molecules. Because molecular vibrations are anharmonic, the transition from the first excited state to the second excited state of the vibrational manifold of a real molecule will occur with an energy that is slightly less than the transition from the ground state to the first excited state. Thus, the increase in absorbance that results from the

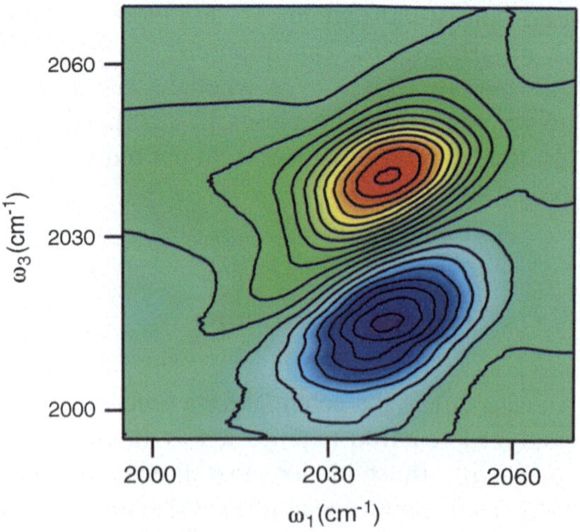

Fig. 2 2D IR spectrum of azide in D_2O at $T_w = 100$ fs illustrating the two types of signals in a typical 2D IR spectrum

population of molecules in the first excited state occurs at a frequency in ω_3 that is below that of the fundamental absorption from the ground state to the first excited state. Furthermore, because the excited-state-absorption signal causes an increase in absorbance, it is observed as a negative peak (decreased transmittance) in the 2D IR spectrum. Thus, the two peaks in the 2D IR spectrum of a single oscillator appear in ω_3 at the frequency of the fundamental absorption, often called the 0–1 frequency, and at the anharmonically shifted excited-state absorption, often called the 1–2 frequency. Because these signals result from vibrational excitation of the molecules, which moves population from the ground state to the first-excited state, the 2D IR signals that give rise to these two peaks will decrease in amplitude as the excited vibrational population relaxes back to the ground state as we increase the time delay between excitation and detection pulses, T_w. In fact, the amplitude of the features in the 2D IR spectrum will decay exponentially as a function of increasing waiting time with a time constant T_1 that reflects the first-order relaxation kinetics of the vibrational excited state. With protein samples, we can, in practice, measure 2D IR spectra for waiting times up to $\sim 5T_1$. As discussed below, analysis of the positive peak (in red) is sufficient to determine the environmental dynamics.

The important information in a 2D IR correlation spectrum is not found in the peak positions or amplitudes, however. Rather, the most useful information is found in the shapes of the peaks in the 2D IR spectrum. To understand the shapes of the peaks we first have to think about the behavior of an ensemble of identical oscillators in solution. In the gas phase, these identical oscillators have identical energy-level spacings and transition frequencies. In solution, however, each molecule experiences a slightly different solvent environment. The interactions with solvent molecules cause a slight shift in the energy levels of the oscillator. Thus, a distribution of identical oscillators in solution exhibits a distribution of transition frequencies. This distribution, however, is not static. The transition frequency for a particular molecule changes as a function of time as the molecule samples different solvent environments. This process by which molecules sample the distribution of frequencies is known as spectral diffusion. In a 2D IR experiment, the frequency distribution produces a line shape that is elongated parallel to the diagonal for short waiting times, T_w. For small enough T_w, a molecule that has a low transition frequency when it is excited in ω_1 also has a low transition frequency when it is probed in ω_3 because the molecule has not yet sampled other environments. The situation is similar for a high frequency oscillator. Thus, there is a strong correlation between the frequency in ω_1 and the frequency in ω_3 for small T_w delays. As a result the lineshapes of the peaks in the 2D IR spectrum are elongated along the diagonal at early waiting times. At longer waiting times, however, the molecules have an opportunity to sample different environments in between the excitation pulse pair and the probe pulse. As a result, the frequency in ω_3 is increasingly less correlated with the frequency in ω_1 as T_w increases and the line shape in the 2D IR spectrum tilts more towards the horizontal and becomes more rounded. Figure 3 shows a series of 2D IR spectra for increasing waiting times from top to bottom illustrating the effect of spectral diffusion as a function of the waiting time on the 2D IR line shapes.

The 2D IR line shape measures frequency correlations as a function of the waiting time. The loss of correlation between the frequencies in ω_1 and ω_3 reflects the spectral diffusion that results from the changes in environments around the oscillators in the sample. Thus, the time scales on which the line shape in the 2D IR spectrum changes are the time scales for the dynamics of the environment around the probe molecules. We quantify these time scales statistically using the frequency–frequency time correlation function (FFCF). Mathematically, the FFCF is defined as $\langle\delta\omega(t)\delta\omega(0)\rangle$ where the brackets indicate the ensemble average, and $\delta\omega(t)$ is defined as $\omega(t) - \langle\omega\rangle$, the instantaneous frequency of an oscillator minus the average of the frequency distribution. At $t = 0$, the FFCF gives the mean square of the frequency fluctuations. As t increases, the FFCF decays reflecting the loss of correlation between the frequency fluctuation at time t and its value at time 0.

The line shape of the 2D IR spectrum as a function of T_w is a quantitative measure of the FFCF. It is not trivial, however, to determine the FFCF uniquely from a series of 2D IR spectra. Among the many approaches to extracting the FFCF from 2D IR correlation spectra, one of the most straightforward and robust is the centerline slope (CLS) method first introduced by Kwak et al. [30, 31]. The CLS is constructed by taking slices of the 2D IR spectrum for specific values of ω_1 and determining the location of the maximum in ω_3 for the slice. Thus, the centerline is the collection of values that correspond to the location of the maximum in ω_3 for each value of ω_1. For the 2D IR spectra shown in Fig. 3, the blue markers show the points corresponding to the centerline, which is fit to a linear function as shown by the green line.

When the correlation between the frequencies of the oscillators in ω_1 and ω_3 is high, the centerline in aligned with the diagonal, and as the frequency correlations decay with increasing waiting times, the centerline rotates toward horizontal following the changes in the 2D IR lineshape. This evolution is characterized by the slope of the centerline, the CLS, which has a theoretical maximum value of 1 if the correlations in ω_1 and ω_3 are perfect. At large enough waiting times, the CLS

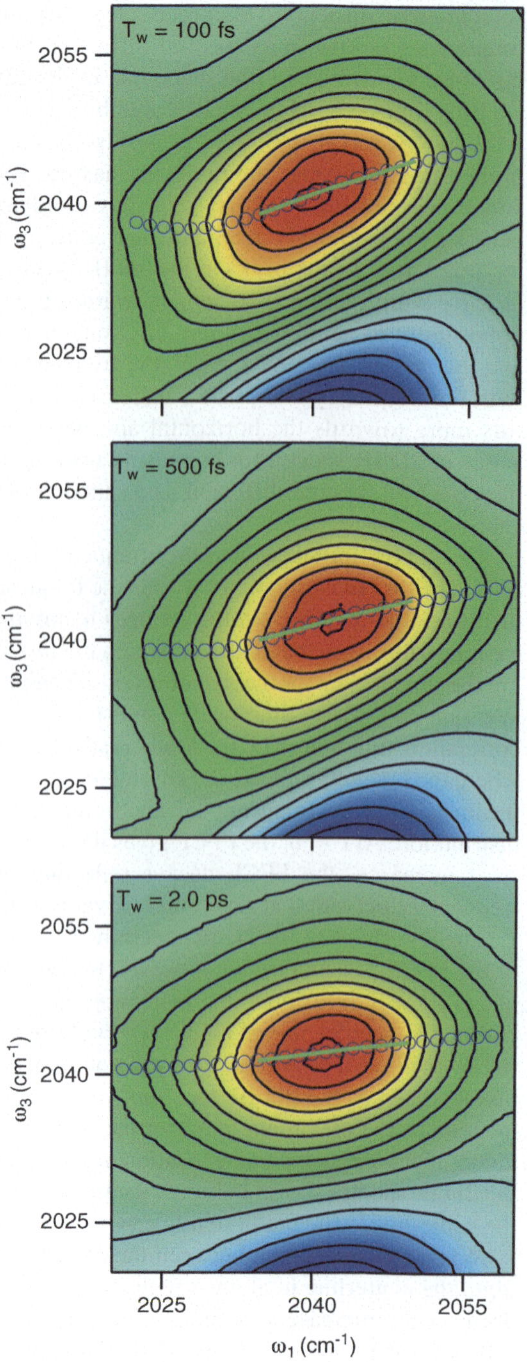

Fig. 3 2D IR spectra of azide in D_2O for waiting times (T_w) of 100 fs, 500 fs, and 2.0 ps illustrating the centerline slope (CLS) analysis of the 2D IR lineshape. The *blue circles* are the points of the centerline and the *green line* is the fit to these data from which we get the CLS

must decay to zero since the frequency correlations are lost completely when the molecules undergo spectral diffusion as a result of sampling the full distribution of environments. The time scales for the decay of the CLS, therefore, are a direct measurement of the time scale of the fluctuations of the environment. In general these time scales are quantitatively characterized from the CLS decay using a generalized Kubo lineshape function in which the decay of the FFCF is assumed to be multiexponential. Thus we fit the decay of the CLS to a sum of exponentials. Using the exponential time scales and their relative amplitudes from the decay of the CLS, we then fit the lineshape of the infrared absorption spectrum to determine the FFCF with the only adjustable parameters being the absolute amplitude of the FFCF and the motionally narrowed contribution to the lineshape. Thus, the FFCF will most often be reported in terms of amplitudes and time scales that correspond to a function of the form

$$C(t) = \Delta_s^2 + \sum_i \Delta_i^2 \, e^{-t/\tau_i}, \tag{4}$$

where the summation runs over as many terms as are necessary to characterize the FFCF decay. The Δ values give the amplitudes of the frequency fluctuations that contribute to the FFCF and the τ values give the time scales for those motions. The static contribution, given by the Δ_s term, is included to account for those motions that occur on time scales that are longer than the measurement time afforded by the limits associated with population relaxation and are therefore not sampled within the accessible waiting times. Examples of fits to experimental data for Eq. (2) are given in below in Sects. 3 and 4.2.

Thus, 2D IR spectra as a function of the waiting time allow us to determine uniquely the time scales of the environmental fluctuations about a probe molecule. In the study of enzyme dynamics, the probe molecule is chosen based on four properties:

1. *Its ability to bind to the enzyme active site.* The molecules must bind to the protein to report on enzyme dynamics, and the most relevant location to probe these motions is in the active site. Furthermore, if the molecule is a transition-state-analog inhibitor, this is an added advantage in that the dynamics that are most relevant to the enzyme-catalyzed reaction will be those that occur in the transition-state complex. If the complex with the probe molecule bound mimics the transition-state complex well, then the measurements will better represent the dynamics of the most chemically relevant state of the enzyme.
2. *The position of the transition frequency in the infrared spectrum.* The probe molecule must also have its infrared transition in a suitable region of the infrared spectrum. Naturally, the solvent in enzyme dynamics experiments will be water, which has strong absorption bands throughout the infrared. The most suitable region for isolating a particular transition in an aqueous enzyme solution is between 1,800 and 2,800 cm^{-1}. There are relatively few potential chromophores that absorb in this region. The most commonly used candidates are triple-bonded small molecules or ions with absorptions associated with the triple-bond stretching motions around 2,000 cm^{-1}, such as carbon monoxide, nitriles, and azides.

3. *The intensity of the infrared spectroscopic transition.* The molecule must also have a strong transition moment for the relevant spectroscopic transition. The signal strength in 2D IR spectroscopy scales as the square of the molar absorptivity. Thus a weak chromophore will be very difficult to detect in 2D IR. This problem is amplified in measurements of enzyme dynamics because the maximum solubility of enzymes is typically on the order of 1 mM or less, which, although it is a large concentration for an enzyme, is a very small concentration for infrared spectroscopy. Consequently, with current technology, it is best if the molar absorptivity of the spectroscopic transition of interest in the probe molecules is on the order of 1,000 M^{-1} cm^{-1} or greater.
4. *The population lifetime of the chromophore.* The chromophore should also have a long population lifetime. As noted above, the population lifetime of the probe vibration sets an upper limit on the spectral diffusion time scales that are accessible by 2D IR spectroscopy. In practice we can measure 2D IR spectra of proteins for waiting times up to five times the population lifetime. Because proteins can exhibit conformation dynamics over many decades of time, a probe molecule with a long vibrational lifetime provides access to dynamics on a much wider range of time scales.

2 Temperature Dependence of KIEs in Enzymes and Solution

Several enzymatic and a few non-enzymatic systems have been examined using TDKIEs. A few examples of these systems are concisely presented here to demonstrate some applications of this methodology. The examples presented below focus on C–H bond activation via hydride or proton transfer (adiabatic processes) that are relevant to the studies of formate dehydrogenase (FDH) discussed in Sect. 4.

2.1 Dihydrofolate Reductase

Dihydrofolate reductase (DHFR) catalyzes a simple chemical transformation (Scheme 1): the NADPH-dependent reduction of 7,8-dihydrofolate (H_2folate) to 5,6,7,8-tetrahydrofolate (H_4folate, Scheme 1). H_4folate is an important cofactor in many biochemical processes including thymine (a DNA nucleotide) biosynthesis, and thus DHFR is a target for various antibiotic and chemotherapeutic drugs. DHFR has served as a platform for many theoretical [2, 3, 32–38] and experimental studies [26, 39–42], and the relationship between protein dynamics and function has been specifically examined in DHFR [43–45]. As a result, DHFR is an ideal model system for exploring basic physical features of enzymology. In spite of these advantages, the chemical step studied by theoreticians, and illustrated in Scheme 1, is not the rate-limiting step of the enzymatic turnover and a lot of work has been

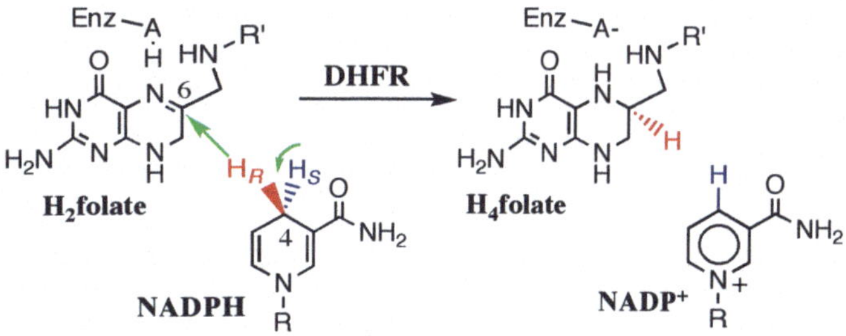

Scheme 1 The DHFR reaction. The *green arrows* indicate the motion of the primary (H_R) and secondary (H_S) hydrogens. R: adenine dinucleotide 2′-phosphate, R′: *p*-aminobenzoyl-glutamate

Table 1 Comparative kinetic parameters of the DHFR I14 mutants

Parameters	WT	I14V	I14A	I14G
Residue volume[a] (Å^3)	124	105	67	48
k_H[b] (s^{-1})	228 ± 8	33.3 ± 3.1	5.7 ± 0.3	0.22 ± 0.04
A_H/A_T[c]	7.0 ± 1.5	4.2 ± 0.4	4.7 ± 0.5	0.024 ± 0.003
ΔE_{aT-H}[c] (kcal/mol)	−0.1 ± 0.2	0.27 ± 0.05	0.39 ± 0.06	3.31 ± 0.07

[a]Side chain volume [50]
[b]Presteady state rates of H transfer at 25°C and pH 7
[c]Similar trends were observed for H/D and D/T (data not shown)

invested in exposing that step and its intrinsic KIEs using the methods described above [41, 46–48].

Escherichia coli DHFR was used to study changes in the C–H → C transfer in response to a perturbation of the DAD. We altered the DAD and its dynamics by reducing the size of an active site isoleucine 14, located behind the H-donor, into valine, alanine, and glycine. These mutations minimize electrostatic changes, and an MD simulation indicated that the mutations primarily affect the distribution of DADs, resulting in longer average values and a broader distribution (Fig. 1c) [49]. Pre-steady-state kinetics indicates slower rates for smaller side-chain derivatives (Table 1), in accordance with a smaller fraction of reactive conformations [C_T in Eq. (3)]. Competitive H/T and D/T KIEs (in the 5–45°C range) afford the intrinsic KIEs and their temperature dependence (Fig. 4), which, together with the Arrhenius analysis, is provided in Table 1.

These findings indicate a clear relationship between the DAD distribution and the temperature dependence of intrinsic KIEs. These local effects also enable a better understanding of remote mutations that indicate synergism between residues >15 Å away from the active site that are associated with a dynamic network of coupled motions across the protein that is coupled to the catalyzed chemistry [41, 46–48]. Although the earlier studies support the idea of a dynamic network suggested by three independent theoretical studies [2, 42, 51], the structural and dynamic effects of the remote mutations on the DAD were hard to assess without similar effects in active-site mutants and the associated MD simulation [49].

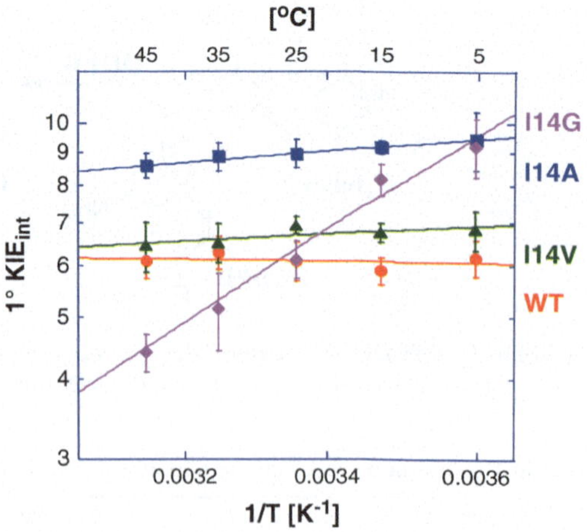

Fig. 4 Arrhenius plot of intrinsic H/T KIEs (on a log scale) for wild-type (*red*), I14V DHFR mutant (*green*), I14A DHFR (*blue*) and I14G DHFR (*purple*). The *lines* represent the nonlinear regression to an exponential equation. Reproduced with permission from [49]

2.2 *Thymidylate Synthase (TSase)*

TSase catalyzes the last step of de novo synthesis of 2′-deoxythymidine-5′-monophosphate (dTMP, one of the four DNA bases), in which the substrate 2′-deoxyuridine-5′-monophosphate (dUMP) is methylated and reduced by the cofactor N^5,N^{10}-methylene-5,6,7,8-tetrahydrofolate (CH_2H_4folate) (Scheme 2). The catalytic mechanism of TSase involves a series of bond cleavages and formations that include two different C–H bond activations: a reversible proton abstraction from C5 of the substrate, dUMP and an irreversible, rate-limiting hydride transfer from C6 of the cofactor to C7 of the product, dTMP. Early mechanistic studies of TSase included steady state and pre-steady state kinetics, crystallography, and 1°- and 2°-hydrogen isotope effects, etc., which have been summarized in a few excellent reviews [52–55]. We studied the two C–H activation steps with both experimental and computational methods, and highlight here the integration of different approaches in addressing specific issues [56–63]. In this section KIEs refer to 1° KIEs and TSase refers to the enzyme from *E. coli*, unless otherwise specified.

1. *The hydride transfer* (step 5 in Scheme 2). Spencer et al. measured KIEs on the hydride transfer with steady state kinetic experiments of wild type TSase (wt TSase) at 20°C [64]. The authors observed a large 1° KIE (~3.7) on both k_{cat} and k_{cat}/K_M, suggesting that the hydride transfer step was rate limiting for the catalysis (i.e., lack of "kinetic complexity"). Subsequently, we measured intrinsic 1° KIEs on the hydride transfer step in the physiological temperature range (5–40°C) of wt TSase [56]. The small kinetic commitment [Eq. (1)] observed on the hydride transfer supports the conclusions of Spencer et al. that this step is rate limiting. Over the experimental temperature range (5–40°C), the intrinsic KIEs are temperature independent (Fig. 5a), and the isotope effects on the Arrhenius pre-exponential factors exceed the upper limit of semi-classical predictions. We also measured initial velocities for the reaction over the same temperature range, and

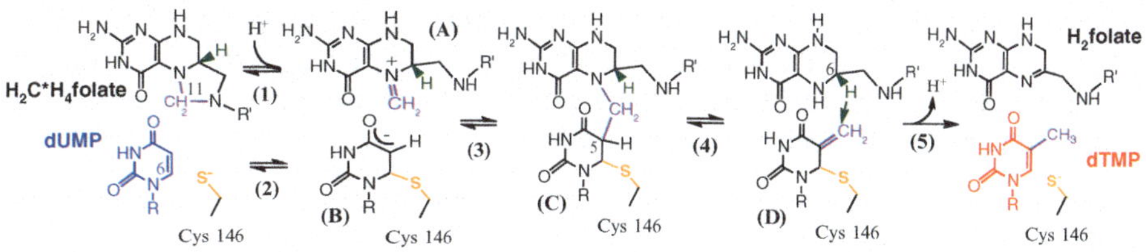

Scheme 2 TSase catalyzed reaction starts with Michael addition of Cys146 (step 2) and methylene transfer from methylenetetrahydrofolate (CH_2H_4folate) to deoxyuridine monophosphate (dUMP) (step 3) followed by proton abstraction from C5 of dUMP and elimination of tetrahydrofolate (step 4), followed by hydride transfer to produce dihydrofolate (H_2folate) and deoxythymidine monophosphate (dTMP) (step 5)

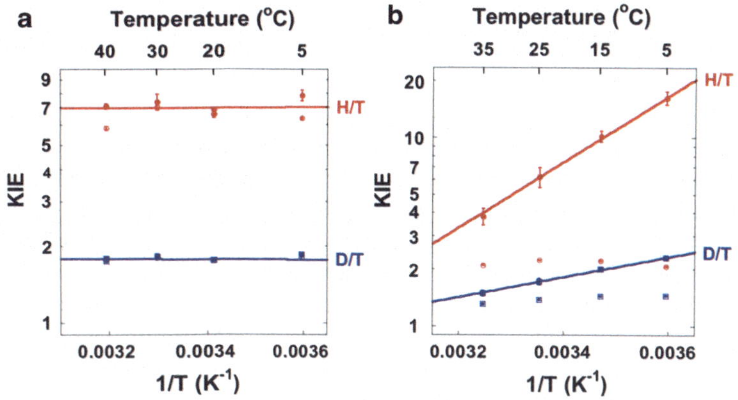

Fig. 5 Arrhenius plots of observed (*open markers*) and intrinsic (*filled markers*) primary KIEs on (**a**) the hydride transfer (step 5 in Scheme 2) and (**b**) the proton transfer (step 4 in Scheme 2) catalyzed by TSase. The hydride transfer (rate-limiting) presents temperature-independent KIEs, while the proton transfer presents temperature-dependent KIEs. This figure is modified from [56, 63]

observed a significant activation energy on k_{cat}. This system is another example of an enzyme-catalyzed H-transfer with a temperature-dependent rate but temperature-independent KIEs, which is consistent with Marcus-like models but not with simple tunneling corrections to TST.

2. *The proton transfer step* (step 4 in Scheme 2). Compared with the rate-limiting hydride transfer, the proton transfer is a fast and reversible step and therefore its KIEs are greatly masked by the kinetic complexity. Previous studies used a saturating concentration of CH_2H_4folate and measured a 1° KIE of unity on the proton transfer due to the large kinetic commitment [55, 65]. We found that the observed KIE for this step varies with the concentration of CH_2H_4folate, owing to the sequential binding order of dUMP and CH_2H_4folate [58]. Thus, we measured the KIEs on the proton transfer with a low concentration of CH_2H_4folate (yet high enough to ensure sufficient conversion of dUMP to dTMP) [63], and intrinsic KIEs were assessed using Northrop's method [Eq. (2)]. In contrast to the hydride

transfer step, the intrinsic KIEs on the proton transfer are temperature-dependent (Fig. 5b). The different temperature dependences of the KIEs on the two hydrogen transfer steps suggest that the enzyme employs different strategies in catalyzing the sequential C–H bond activations [63].

This is a unique example of two different C–H activations being examined in the same catalytic conversion and same active site. These findings indicate that the enzyme has evolved to tune carefully the rate limiting step (which does not happen in solution, suggesting an enormous catalytic effect), but that there is little intervention of the enzyme in the fast proton abstraction (which can be catalyzed in solution by high pH and thiols).

2.3 *Aromatic Amine Dehydrogenase and Morphinone Reductase*

Scrutton, Sutcliffe, and coworkers have employed the temperature dependence of intrinsic KIEs to study several enzymatic systems. We discuss only one such study, for which the scope of methodologies ranged from kinetics to QM/MM examination of two enzymes. Masgrau et al. [66] examined the reaction pathway for tryptamine oxidation by aromatic amine dehydrogenase using a combined experimental and theoretical approach. They were able to demonstrate that proton transfer in this system is dominated not by long-range promoting vibrations but by short-range motions that modulate the distance between the proton and the acceptor oxygen atom. Similarly, Hay et al. [67] used both temperature and pressure dependence measurements to study the reductive half-reaction of morphinone reductase (MR). This reaction has been proposed to involve hydride tunneling from NADH to the enzyme-bound cofactor flavin mononucleotide (FMN) [68], and the measurement of the pressure-dependence of the 1° KIE for this system allowed the authors to suggest a full tunneling model for the hydride transfer that authors describe as an environmentally coupled tunneling model [21], addressed here as an example of a Markus-like model.

2.4 *Thermophilic Alcohol Dehydrogenase from* **Bacillus stearathermophilus** *(ht-ADH)*

For this enzyme, studies indicate that the H-transfer is rate limiting across the temperature range studied. ht-ADH exhibits temperature-independent 1° KIEs within its physiological temperature range (30–65°C) despite the fact that the overall rate is temperature-dependent [69]. As noted above, temperature-independent KIEs with a temperature-dependent rate can be rationalized by the Marcus-like models described, where isotopically insensitive heavy atom motions compose the reaction coordinate, or as stated by Kiefer and Hynes "the solvent coordinate is the reaction coordinate"

[70]. Interestingly, below its physiological temperature range, ht-ADH shows temperature-dependent KIEs. This indicates a sort of phase transition at low temperatures, which alters the DAD sampling at the TRS (Fig. 1c). This phenomenon led to a proposal that certain protein motions bring the donor and acceptor close enough for hydrogen tunneling, but those motions are constrained below the physiological temperature range. Recent studies of mutants of this enzyme have further supported and extended this interpretation [71].

2.5 *Non-enzymatic Systems*

Both temperature-dependent and independent KIEs have been measured for several non-enzymatic systems. Several C–H activations in intramolecular reactions that exhibit temperature-independent KIEs are especially of interest in connection with the enzymatic systems described above [72]. While no rigorous explanation was proposed at the time, these results suggest that such systems could be used to understand biological catalysis better. More recent solid state NMR studies of intramolecular N–H $\rightarrow$ N reactions have followed a more semiclassical path, where KIEs are temperature-dependent at high temperature and become temperature-independent at low temperature [73].

3 2D IR Spectroscopy of Proteins

The dynamics of several different proteins have been examined by 2D IR spectroscopy. We summarize the results of a few such studies here to demonstrate the application of 2D IR to these systems and to identify common features and patterns that emerge in the dynamics of proteins. We have organized this summary by the type of chromophore that has been used to probe the protein spectral diffusion as the chromophore significantly impacts the range of experimentally accessible spectral diffusion time scales and the interpretation of the dynamics. The first few examples describe experiments that employ carbon monoxide as the probe. Carbon monoxide is a strong vibrational chromophore that binds tightly to the iron center of heme proteins. The CO stretching vibration absorbs at approximately 1,950 cm^{-1}, which is an accessible region of the infrared spectrum with little interference from protein or water vibrations. These characteristics make CO an excellent probe for a wide range of heme proteins [74–85]. The last few examples describe experiments that use cyano [86] or azido [87–91] vibrations to probe the dynamics of several different enzymes. Cyano groups, which absorb near 2,200 cm^{-1}, have great potential as probes of biomolecular dynamics [92], but have seen limited applications because of their relatively weak transition dipole moment. Azido groups exhibit much stronger transition dipoles for the antisymmetric stretching vibration that absorbs at 2,040 cm^{-1} in the azide anion and at somewhat higher frequencies for azido-derivatized compounds. Azides have seen more applications to protein dynamics because they can be incorporated into a wider range of systems than carbonyls, but suffer from relatively short population lifetimes that limit the time scales over which 2D IR can observe the dynamics.

3.1 Effect of Denaturation

As a protein denatures, the relatively narrow conformational distribution of the native protein gives way to a complex conformational ensemble characteristic of a rugged energy landscape with many local minima. The dynamics of such an ensemble are sure to be different from those of the native complex. At the same time, however, the probe molecule is likely to become solvent exposed, potentially resulting in spectral diffusion that is more characteristic of the solvent than of the ensemble of protein conformers. Kim et al. studied the effect of protein denaturation on the observed spectral diffusion dynamics for CO bound to the M61A mutant of cytochrome c from *Hydrogenobacter thermophilus* [78]. Table 2 shows the FFCF parameters for CO bound to the protein both in normal aqueous buffer solution and at 5.1 M concentration of guanidinium HCl (GuHCl). In the native state the protein exhibits dynamics on two time scales: an 8 ps component and a component that occurs on a time scale longer than the approximately 100 ps measurement time scale. Denaturing the protein with guanidinium slightly increases the time scale of the fast component, but the more significant effect is that the amplitude of the frequency fluctuations at both time scales increases, with a dramatic increase in the amplitude of the static component. This static contribution to the FFCF suggests that there is a broad distribution of structures present even in the native protein, and that the heterogeneity of the ensemble increases significantly upon denaturation, as would be expected. Although the dynamics of water are very fast, typically inducing nearly complete decay of the FFCF of an aqueous probe vibration in 1 ps or less [93–98], the CO group in the denatured protein does not exhibit dynamics that reflect interactions with bulk water. Instead, the large static contribution to the FFCF suggests that the denatured state at 5.1 M GuHCl is a heterogeneous ensemble of protein conformations that interconvert only on very long time scales and that interactions within these protein conformations give rise to the inhomogeneous distribution of CO transition frequencies. It is likely that the static contribution to the FFCF in these experiments is characteristic of the denatured ensemble and may, therefore, be different if the protein were to be denatured by a different mechanism. Chung et al. studied the dynamics of the same enzyme as a function of temperature both below and above the thermal denaturation temperature [76]. Below the denaturation temperature, the FFCF parameters are nearly temperature independent. Above the unfolding transition, however, all of the FFCF parameters become temperature dependent. The amplitude of the fluctuations that relax on the picosecond time scale, Δ_1, is larger for the unfolded state that for the native value by a factor of nearly 2 and continues to increase with temperature. The corresponding FFCF decay time, τ_1, for the unfolded state is also larger by a

Table 2 FFCF parameters for native and denatured cytochrome *c* [78]

Protein	Δ_1 (ps^{-1})	τ_1 (ps)	Δ_s (ps^{-1})	T_1 (ps)
Cyt-*c*	0.79	8	0.74	29
Cyt-*c* (5.1 M GuHCl)	0.89	10	1.9	23

factor of nearly 2 than for the folded state but decreases with temperature. Finally, the static contribution to the FFCF, Δ_s, increases almost threefold relative to the folded protein and also decreases with temperature. These trends have a straightforward interpretation in terms of the protein dynamics. The decrease of Δ_s with increasing temperature reflects the fact that, as the temperature increases, more of the protein conformational fluctuations occur within the ~100 ps measurement time. This trend is consistent with the temperature dependent rise in Δ_1. Thus, the unfolded state exhibits much more heterogeneity than the folded state regardless of whether the denaturation is thermal or chemical. In addition, as the temperature increases, the unfolded protein samples the available conformational space increasingly rapidly. Thus, these results suggest that the unfolded state represent a rugged conformational landscape with many local minima and that the increasing temperature increases the rate of crossing the kinetic barriers between these minima, leading to the increased rate of sampling the conformational distribution.

3.2 *His-Tag Perturbations*

Histidine tags (His_6) are frequently used to simplify protein purification and are assumed to make only minor perturbations to the protein structure or function. Thielges et al. have studied the effect of a His tag on the spectral diffusion dynamics of CO bound to myoglobin (MbCO) [75]. Table 3 shows the FFCF parameters for MbCO and MbCO-His_6. The presence of the His tag eliminates the contribution to the FFCF from the short time scale motions and slightly increases the static contribution to the FFCF. The effect of the His tag on the observed dynamics is surprising because there is no accompanying structural perturbation. In addition, the His tag is a relatively small collection of residues added to the end of the protein which is located far from the heme pocket. Nevertheless, it is clear from this study that the presence of the His tag does affect the protein dynamics at the active site. Although this effect is modest, the fact that it is measurably large is somewhat surprising and underscores the potential for distal perturbations to impact the dynamics at the functional binding site of a protein.

3.3 *Disulfide Bond Effects*

Disulfide bonds contribute to protein stability and help to regulate protein activity. It seems likely that they should also have a significant impact on the conformational

Table 3 FFCF parameters for CO bound to myoglobin and myoglobin with a His tag [75]

Protein	Δ_1 (ps^{-1})	τ_1 (ps)	Δ_2 (ps^{-1})	τ_2 (ps)	Δ_s (ps^{-1})	T_1 (ps)
MbCO	0.23	1.4	0.51	19	0.43	17
MbCo-His_6	–	–	0.51	21	0.57	20

Table 4 FFCF parameters for CO bound to wild type neuroglobin (wt-Ngb), neuroglobin in which the disulfide bond has been reduced (red-Ngb), and neuroglobin in which one of the cysteines involved in the disulfide bond has been mutated to serine (3cs-Ngb)

Protein	Δ_1 (ps^{-1})	τ_1 (ps)	Δ_2 (ps^{-1})	τ_2 (ps)	Δ_s (ps^{-1})	T_1 (ps)
wt-Ngb/N_3	0.36	2.0	0.51	14	0.58	19
red-Ngb/N_3	0.57	1.3	0.62	22	0.55	16
3cs-Ngb/N_3	0.38	0.7	0.57	23	0.43	16
wt-Ngb/N_0	0.34	11.5	–	–	0.66	18
red-Ngb/N_0	0.57	3.7	–	–	0.81	16
3cs-Ngb/N_0	0.53	8.1	–	–	1.1	16

N_3 and N_0 refer to the two conformational substates of the protein [80]

sampling of a protein. Ishikawa et al. reported on the effect of the disulfide bond on the protein dynamics in neuroglobin (Ngb), an oxygen-carrying member of the globin family of proteins that is expressed in the brain and other nerve tissues [80]. Ngb is structurally similar to myoglobin except that it exhibits a disulfide bond. Like in myoglobin, when CO binds to the heme in Ngb, it does so in two distinct conformational substates referred to as N_0 and N_3, which give rise to distinct spectral features for the CO. Table 4 shows the FFCF parameters for these two substates in each of three different forms of Ngb: the wild-type protein (wt-Ngb), the protein with the disulfide bond eliminated by reduction (red-Ngb), and the protein with the disulfide bond eliminated by mutating one of the cysteine residues to serine (3cs-Ngb). These parameters show that the trends are similar for both conformational substates on the short time scale. In each case, elimination of the disulfide bond either by reduction or by mutation accelerates the sampling on the few picosecond time scale and increases the amplitude of these frequency fluctuations, but this effect is less dramatic for the N_0 state of the cysteine mutant protein. On longer time scales the two substates show distinct dynamics and trends with respect to disulfide bond cleavage. For N_3, the static component gets somewhat smaller as the time scale for the intermediate time component gets somewhat larger. These two observations suggest that some of the motions that contribute to the static component in the wild type are accelerated when the disulfide bond is broken so that they now occur within the measurement window of the 2D IR experiment. For N_0, however, the effects are subtler. The time scale for the spectral diffusion component of the FFCF gets faster upon reduction of the disulfide bond. Mutation of the cysteine accelerates the fast dynamics as well but not as much as in the case of reduction. In addition, disulfide bond cleavage slightly increases the size of the static contribution to the FFCF, suggesting that there is a larger range of accessible conformations for the N_0 substate without the disulfide bond than there is in the wild-type enzyme. Perhaps the most critical feature of this study is that the disulfide bond is located nearly 20 Å from the binding pocket of the protein, yet there are perturbations to the dynamics of both substates of the protein which involve distinct conformations of the distal histidine. Thus, the changes in the dynamics associated with elimination of the disulfide bond must result from global changes in the protein dynamics rather than being more localized perturbations of the dynamics of the CO binding pocket.

3.4 *Horseradish Peroxidase*

Horseradish peroxidase (HRP) is an iron-heme enzyme that catalyzes the oxidation of a wide variety of organic substrates by hydrogen peroxide. Finkelstein et al. report a 2D IR study of HRP with CO bound to the heme in which the dynamics of the CO frequency fluctuations for ternary complexes with several different substrate analogs are compared to the dynamics of the binary complex without a substrate or substrate analog bound [83]. Table 5 shows the FFCF parameters for the binary and ternary complexes of HRP. The binary complex has two substates (red and blue) that correspond to the two protonation states of the distal histidine in the active site. The red-shifted substate with the histidine protonated is the only state present with the substrates bound. In the binary complex this substate exhibits dynamics on two time scales: a fast component with a time constant of 1.5 ps and a slow component with a time constant of 21 ps. All of the ternary complexes exhibit qualitatively similar dynamics: the fast time constant is slower by a factor of 2 or 3 compared to the binary complex and the slow component becomes too slow to observe on the measurement time scale and so becomes a static component to the FFCF decay. In addition to the overall slowing of the dynamics, however, the most striking feature of the dynamics of the ternary complexes is that the amplitudes of both components of the FFCF decay decrease relative to the binary complex, and, in the case of the slower component, which becomes static in the ternary complexes, the amplitude decreases by a factor of between 2 and 3. This effect is even more significant given that the contribution to the FFCF for each component is proportional to the square of the amplitude meaning that this contribution to the FFCF decay is between four and nine times smaller in the ternary complexes than in the binary complex. As a result, the infrared absorption lineshape for all of the ternary complexes is much narrower than that for the binary complex. These differences suggest that forming the ternary complex greatly diminishes the substantial dynamics of the distal histidine and arginine, which are expected to have the greatest influence on the CO frequency fluctuations. Although these dynamics are not fully arrested, the amplitude of the structure fluctuations is substantially reduced and the barriers to such motions become much larger such that the time scales for the remaining dynamics slow considerably. Thus the active site becomes structurally constrained in the ternary complex.

Table 5 FFCF parameters for CO bound to horseradish peroxidase

Protein	Δ_1 (ps^{-1})	τ_1 (ps)	Δ_2 (ps^{-1})	τ_2 (ps)	Δ_s (ps^{-1})	T_1 (ps)
Binary blue	0.60	15.0	–	–	0.45	12
Binary red	0.58	1.5	1.1	21	–	8
2-Napthohydroxamic acid	0.60	2.6	–	–	0.34	–
Benzhydroxamic acid	0.43	4.4	–	–	0.36	–
Benzamide	0.53	4.5	–	–	0.51	–
Benzhydrazide	0.49	2.6	–	–	0.51	–
N-Methylbenzamide	0.32	5.4	–	–	0.38	–

The binary complex of HRP with CO has two substates denoted red and blue. The substrate-analog compounds identify the ternary complexes

3.5 *Cytochrome P450*

The cytochrome P450s are a class of enzymes that hydroxylate a variety of substrates. Thielges et al. report the dynamics of the camphor specific cytochrome P450 (cyt $P450_{cam}$) from *Pseudomonas putida* using CO bound at the active site as a probe [74]. Table 6 gives the FFCF parameters for the cyt $P450_{cam}$ complexes.

The binary complex of the enzyme, with the CO but without a substrate molecule bound, exhibits three distinct features in the CO absorption spectrum reflecting three distinct conformational substates. 2D IR spectra show that these conformers exhibit distinct dynamics on the picosecond time scale and that substrate binding selectively stabilizes a single conformational substate that exhibits dynamics distinct from those of the substrate-free enzyme. Although all three conformational substates of the binary complex exhibit a single picosecond decay component and a static contribution to the FFCF, the FFCF decays for all of the ternary complexes are biexponential with a fast time scale component that is faster than the picosecond decay for any of the substates of the binary complex and a slow component that replaces the static component from the binary complex and relaxes on the time scale of hundreds of picoseconds. These results indicate that, although substrate binding involves some degree of conformational selection, the dynamics of the substrate-bound conformer are significantly influenced by the presence of the substrate invoking features of an induced-fit binding mechanism. By studying complexes of the natural substrate, camphor, and a series of related substrates, the authors identify a correlation between the active site dynamics with each substrate and the binding affinity. Specifically, substrates with smaller K_d values show longer time constants for the slow component of the FFCF decay. The authors conclude that the conformational dynamics of the enzyme becomes constrained in the more tightly bound complexes because restricting the conformational fluctuations of the active site as the system approaches the transition state structure helps to ensure that the hydroxylation reaction occurs at the appropriate carbon. Thus, the specificity of the hydroxylation is related to the rigidity of the active site, supporting the hypothesis that the protein dynamics play a functional role in controlling the outcome of the catalyzed reaction.

Table 6 FFCF parameters for CO bound to cyt $P450_{cam}$

Protein	Δ_1 (ps^{-1})	τ_1 (ps)	Δ_2 (ps^{-1})	τ_2 (ps)	Δ_s (ps^{-1})	T_1 (ps)	K_D (μM)
Binary 1,937 cm^{-1}	0.62	17.0	–	–	1.0	18	
Binary 1,952 cm^{-1}	0.36	9.8	–	–	1.2	21	
Binary 1,963 cm^{-1}	0.32	11	–	–	0.55	24	
Camphor	0.53	6.8	1.0	370	–	19	0.8
Camphane	0.34	5.5	0.72	300	–	27	1.1
Adamantane	0.43	1.6	0.77	260	–	24	50
Norcamphor	0.34	5.8	0.60	110	–	27	345
Norbornane	0.45	2.2	0.89	230	–	23	47

The binary complex of HRP with CO has two substates denoted red and blue. The substrate-analog compounds identify the ternary complexes

3.6 *HIV-1 Reverse Transcriptase*

Reverse transcriptases help to convert single-stranded RNA into double-stranded DNA and are required for the replication of viruses. As a result, HIV-I reverse transcriptase (HIV-I RT) inhibitors are of great interest in the treatment of HIV. Because of the potential for a flexible inhibitor to accommodate mutation-induced changes to the binding pocket, inhibitors with strategic flexibility have the potential to overcome the effects of drug-resistance mutations. Fang et al. report on the dynamics of HIV-1 RT with one such inhibitor TMC278 [86]. A unique feature of this study is that it employs intrinsic vibrations of the inhibitor itself to probe the protein dynamics. In this case, the molecule TMC278 incorporates both a cyanovinyl and a benzonitrile group, each located on one arm of the molecule. These two cyano groups provide spectroscopic handles by which the dynamics of the protein can be observed using 2D IR. This experiment represents a significant challenge to the sensitivity of 2D IR, however, as the protein sample is soluble up to ~1 mM concentration and the nitrile stretching vibration is much weaker than any other chromophores that have been used either before or since for 2D IR experiments on proteins. Nevertheless, the authors are able to measure 2D IR spectra of the complex of HIV-1 RT with TMC278 showing two distinct features corresponding to the two cyano groups of the inhibitor. Unfortunately, the transition moment of the benzonitrile group is small enough for it not to be possible to determine the FFCF for this transition. Nevertheless, the authors do report the FFCF for the cyanovinyl group of the inhibitor but, because of the relatively short lifetime of the CN vibration, they can only measure 2D IR spectra for waiting times up to 5 ps, leaving considerable uncertainty over the long time scale of the decay. The authors report this time scale to be 7.1 ps, and suggest that complete sampling of the nitrile frequency distribution could be complete on the tens of picoseconds time scale. Based on the structure of the binding pocket and the fact that the electrostatic interactions with the rest of the protein are likely to be screened very effectively by the aromatic residues lining the binding site for the cyanovinal group, the authors argue that the spectral diffusion dynamics are dominated by local fluctuations of the cyanovinyl arm with respect to the side chains composing the hydrophobic tunnel. The very local behavior of these dynamics is in contrast to what has been seen with CO in other proteins and reflects one of the impacts of the choice of chromophore on the observed dynamics.

3.7 Carbonic Anhydrase II

Carbonic anhydrases are zinc enzymes that catalyze the interconversion of bicarbonate and carbon dioxide. There are many isozymes of carbonic anhydrase and these are found throughout biology. The most common isozyme in mammals is carbonic anhydrase II (CA II), which has very high sequence homology across species. Lim et al. first reported the dynamics of azide anion bound to the active site zinc of bovine CA II [87]. Subsequently, we studied the dynamics of both wild-type human CA II and two active-site mutants, T199A and L198F [91]. Azide anion is a metal poison and binds to the active site zinc inhibiting the normal activity of the enzyme. Table 7 summarizes the FFCF results for both the bovine CA II (BCA II) and the human CA II (HCA II) as well as the mutants.

For BCA II, Lim et al. report a two-component FFCF decay with a subpicosecond fast contribution and a 17 ps slow contribution. For the wild type HCA II we report a similar fast component but, instead of a slow decay, we report a static contribution to the overall decay. Using the parameters from Lim et al. we get a nearly indistinguishable fit to our data and conclude that these two enzymes exhibit essentially the same dynamics and that the differences in the FFCF reflect differences in how the two groups choose to model the data. In their study of BCA II, Lim et al. suggest that Thr199, the residue closest to the azide in the active site, is likely to be important for modulating the transition frequency of the zinc-bound ligand. This conclusion seems reasonable as the oxygen atom of Thr199 is positioned very near the central nitrogen of the azide, suggesting a strong electrostatic interaction. Based on our experiments on the T199A mutant, however, we can clearly conclude that this residue has almost no effect on the observed dynamics because the FFCF for the T199A mutant are essentially identical to those for the wild-type enzyme. In the L198F mutant, however, we see a significant change to the long-time dynamics. Specifically, the static component that is present in the wild-type enzyme gives way to a 2.5 ps decay. We propose that the phenylalanine ring itself does not directly interact with the azide causing this perturbation. Instead, we suggest that the Phe ring introduces steric hindrance in the active site that forces the azide anion to come closer to the amide proton at position 199 with which it forms a hydrogen bond. This hydrogen bond makes the dominant contribution to the observed dynamics. In the wild-type enzyme the azide anion is free to explore the hydrophobic pocket of the active site and to sample a wide range of hydrogen-bond distances. In the L198F mutant, however, the Phe ring fills the empty space in the active site, significantly constraining the range of hydrogen-bond distances available to the azide anion. As a result, the distribution of

Table 7 FFCF parameters for azide bound to the active site zinc bovine [87] (BCA II) and human [91] (HCA II) carbonic anhydrase II and the human CA II mutants T199A and L198F

Protein	Δ_1 (ps^{-1})	τ_1 (ps)	Δ_2 (ps^{-1})	τ_2 (ps)	Δ_s (ps^{-1})	T_1 (ps)
BCA II	1.6	0.2	0.95	17	–	2.8
HCA II	2.02	0.45	–	–	0.86	–
T199A	1.86	0.40	–	–	0.88	–
L198F	1.96	0.25	1.01	2.5	–	–

hydrogen-bond lengths is sampled much more rapidly, resulting in the 2.5 ps decay contribution to the FFCF.

Common themes from these previous studies of enzyme dynamics are the rigidification of the protein upon ligand binding (proportional to K_d) and the long range of these effects on the protein dynamics. These findings are in accordance with similar trends of rigidification of the protein upon ligand binding on nanosecond to microsecond (ns–μs) time scales measured via NMR relaxation studies of other proteins [40, 99, 100]. While such effects are expected on longer time scales (as suggested by the "induced fit" model of substrate binding), it is less intuitive that such behavior should also be seen on the femtosecond to picosecond time scale.

4 FDH: Combining the Kinetic and Spectroscopic Methods

Formate dehydrogenase (*FDH*) catalyzes the NAD^+-dependent oxidation of formate to CO_2 (Scheme 3). Azide (N_3^-), which is an excellent vibrational chromophore for IR spectroscopy, is a tight-binding inhibitor for FDH. Azide is a potent transition-state analog (TSA) of the catalyzed reaction. Thus FDH can serve to study the relationship between the H-transfer step and the TRS dynamics.

Here we will emphasize two aspects of the examination of FDH: kinetics as a measure of the nature of H-transfer and 2D IR spectroscopy as a measure of active-site dynamics. Finding a correlation between results of these two experiments is the long term goal of the studies described here.

4.1 Kinetics and the Nature of H-Transfer in FDH

In the case of FDH from *Candida boidinii* we had to develop a new method to measure intrinsic KIEs, as labeling of the substrate with both ^{14}C and the isotope of interest (H or D) would have been affected by the 1° ^{14}C KIE. Instead we used [Ad-^{14}C]NAD^+ and formic acid labeled with H or D and trace-labeling of T [101]. This method proved instrumental in measuring competitive H/T and D/T 1° KIEs and the Northrop method was then used to assess the intrinsic KIEs. For the wild-type enzyme these KIEs were temperature independent (Fig. 6).

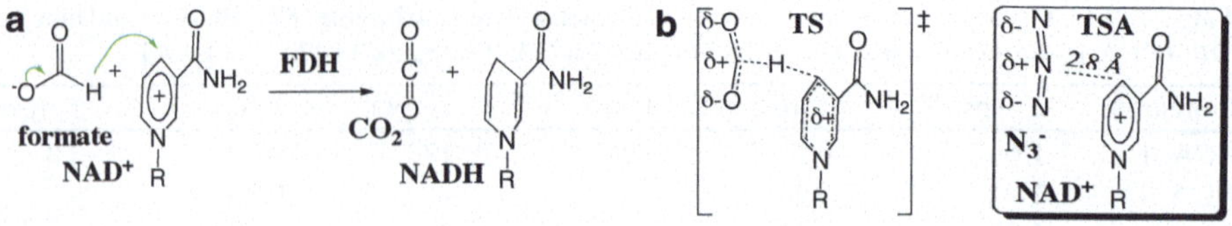

Scheme 3 FDH catalyzed reaction. (**a**) The reaction. (**b**) A comparison of the TS and TSA complexes

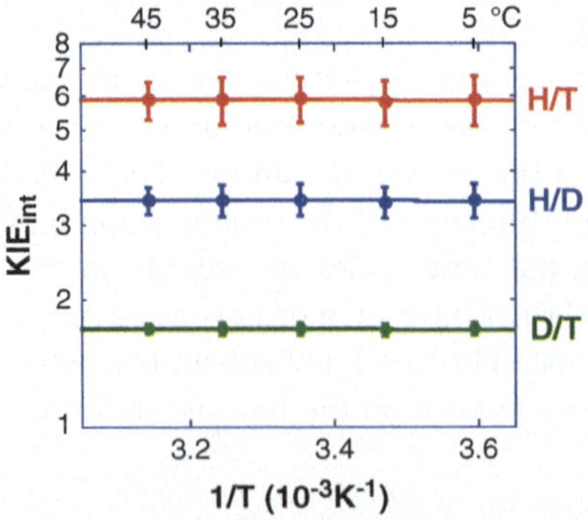

Fig. 6 An Arrhenius plot of the intrinsic H/T (*red*), H/D (*blue*), and D/T (*green*) KIEs (log scale) vs the reciprocal of the absolute temperature. The average KIEs are presented as *points* and the *lines* are an exponential fit of all the data points to the Arrhenius equation

4.2 Dynamics of Transition-State-Analog Complexes of FDH

The ternary complex of FDH with NAD^+ and azide is unlike other protein complexes that have been studied previously in that the azide anion is a transition-state-analog inhibitor. Azide is isoelectronic with the carbon dioxide product of the reaction but is negatively charged like the formate reactant. It is symmetric and acts as a hydrogen-bond acceptor for the highly conserved hydrogen-bond-donating residues in the active site that anchor and orient formate for the reaction. Azide binds tightly in the active site with a K_D of 40 nM. Thus azide is a good mimic of the TRS of the hydride transfer reaction. We have measured the protein dynamics in this complex to explore the potential for femtosecond to picosecond time scale motions to play a functional role in the enzyme-catalyzed hydride transfer [89].

Figure 7 shows the CLS decays for the ternary and binary complexes of FDH with azide. The top panel shows the decay for the ternary complex with FDH, NAD^+, and azide. The center panel shows the decay for the ternary complex with NADH, and the bottom panel shows the decay for the binary complex with just FDH and azide. Table 8 shows the FFCF parameters that result from fitting the decays and the infrared absorption spectra that are shown as insets in Fig. 7.

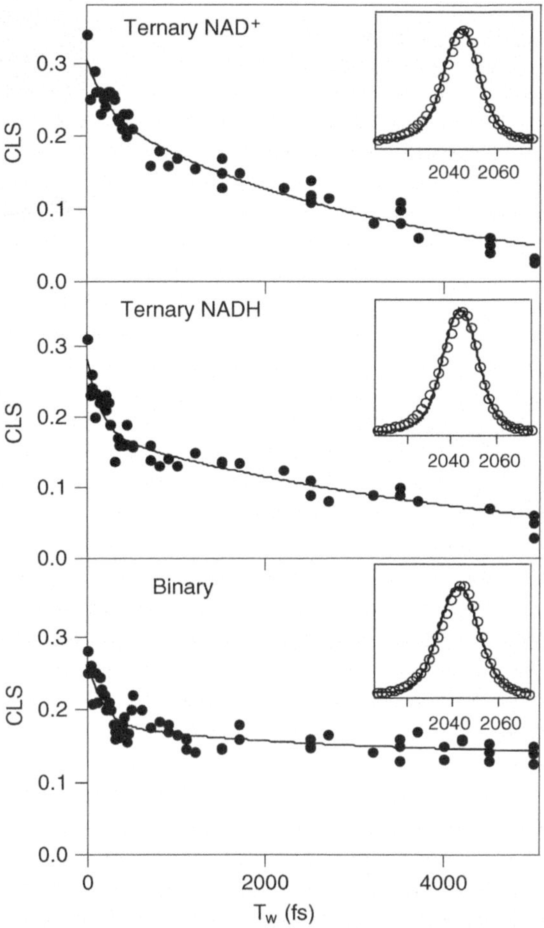

Fig. 7 Decays of the CLS as a function of the waiting time, T_w, for the complexes of FDH with azide. The *lines* represent fits of Eq. (4) to the experimental data (*markers*). The time constants and relative amplitudes are used to fit the FTIR spectrum (*inset*) to get the final FFCF parameters, which are presented in Table 8. Reproduced with permission from [89]

Table 8 FFCF parameters for azide in the ternary and binary complexes with FDH [89]

Protein	Δ_1 (ps^{-1})	τ_1 (ps)	Δ_2 (ps^{-1})	τ_2 (ps)	Δ_s (ps^{-1})
FDH–N_3–NAD^+	0.9	0.21	1.4	3.2	–
FDH–N_3–NADH	1.0	0.15	1.3	4.6	–
FDH–N_3	1.1	0.16	0.8	2.6	1.4

The first thing to note is that there is no static component to the decay for either of the ternary complexes. This behavior is in stark contrast to most other systems that have been studied previously. In the two other cases for which the protein dynamics are fully sampled on the time scale of a few picoseconds, in HIV-1 reverse transcriptase and in the L198F mutant of human carbonic anhydrase II, the result is easily explained in terms of the structure because it can be argued that the dynamics of the probe vibration are dominated by local interactions and do not reflect the global protein dynamics. In other words, because of the nature of the local interactions, the

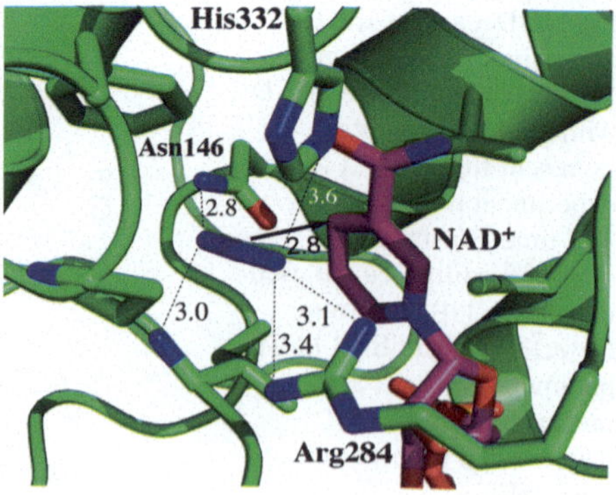

Fig. 8 Active-site structure of FDH–azide–NAD^+ complex (PDB# 2NAD [102]). Azide is in *blue* and NAD^+ in *magenta*. Reproduced with permission from [89]

probe is insensitive to the motions of the rest of the protein. That explanation is not reasonable for the ternary complexes of FDH. Figure 8 shows the active site structure of the ternary complex of FDH with azide and NAD^+ from *Pseudomonas sp.101* [102], which has high sequence homology with the *C. boidinii* enzyme used in our experiments. The critical features to note in this structure are the four hydrogen-bond-donating residues that anchor the azide in the active site: Asn-146, Arg-284, His-332, and the amide N–H of Ile-122. These residues are highly conserved across many species of FDH and are critical to the function of this enzyme because they bind and orient the substrate for reaction. In the context of our spectroscopic measurements, these residues are important because the hydrogen bonds that they donate to the azide anion are likely to be the dominant source of frequency fluctuations and, therefore, play a prominent role in determining the observed FFCF decay dynamics. On each side of the active site is a β-sheet–α-helix–β-sheet sandwich structural motif. Each of these residues resides at the end of one of the major secondary structural elements that comprise this motif. Asn-146 is at the end of an α-helix on one side of the azide. Arg-284 is at the end of a helix on the other side of the azide. His-332 is in the loop region between two strands of the β-sheet, and the amide group of Ile-122 is in a loop region between the end of a helix and the start of a β-sheet strand. Thus, if any of these large secondary structural elements were to undergo the kind of large amplitude motions that would be expected to the be responsible for the long time scale structural dynamics, then we would expect these motions to modulate the hydrogen-bond length between the appropriate residue and the azide in the active site, thereby modulating the frequency of the anion. These motions would be expected, on our measurement time scale, to make a static contribution to the FFCF. In other words, based on the structure, we would expect the azide to be particularly sensitive to the slow motions of the overall protein structure because of the location of the hydrogen bond partners. Nevertheless, we do not observe such a static contribution to the FFCF in the ternary complexes. We conclude, therefore, that there are no dynamics of the active site residues of the protein on time scales longer than a few picoseconds in the transition-state-analog complexes of FDH.

This result seems rather startling. The idea that a protein complex does not exhibit dynamics at time scales greater than a few picoseconds is difficult to reconcile with the fact that proteins have such a large and hierarchical structure. For most protein complexes this structure naturally results in dynamics that occur over a broad range of time scales from femtoseconds to milliseconds. This idea is supported by previous studies of protein dynamics by 2D IR that show a substantial contribution to the FFCF decay on time scales longer than the measurement time scale. What is unusual about the ternary complexes of FDH with azide is that these complexes mimic the TRS for the hydride transfer. If we imagine the potential energy surface for a transition-state complex, it should have a saddle point, i.e., a maximum of energy in one degree of freedom, the reaction coordinate, and a minimum of energy in all other degrees of freedom. Because enzymes are optimized by evolutionary pressure to reduce the barrier in the unbound degree of freedom (i.e., the reaction coordinate) to catalyze the reaction, the potential energy surface in the bound degrees of freedom (orthogonal to the reaction coordinate) should be a single deep potential well that results in a narrow conformational distribution. In the transition-state-analog complex, we prepare the system in a state that takes advantage of all of the stabilizing interactions that are present in the transition state, but this complex does not include the unbound degree of freedom. There is no hydride to transfer, so there is no unbound coordinate. Thus, the potential energy surface of the transition-state-analog complex should have a single deep potential minimum at the TRS configuration. The thermally accessible conformational space in this deep potential minimum should be relatively narrow, and the fluctuations about the equilibrium structure can be fully sampled on short time scales. Thus, in the TRS, we might well expect to see the complete decay of the FFCF within a few picoseconds as in our experiments.

If this interpretation of our experimental results for the ternary complexes is correct, then we predict that the binary complex should behave qualitatively differently because it lacks the coenzyme and does not mimic the TRS. As can be seen in Fig. 7, although the CLSs decay completely on the picosecond time scale for both ternary complexes, in the binary complex the CLS decays to a substantial static offset. This static contribution to the decay reflects all of the motions on time scales beyond the measurement time scale. Thus, this result supports the above interpretation of the data for the ternary complexes. This finding is also in accordance with the results of the kinetic studies in Sect. 4.1, i.e., the orientation of the donor and acceptor at the TRS is optimized for H-tunneling, a condition that requires intimate and strong dynamic restrictions to eliminate all other conformations where the donor–acceptor orientation is non-ideal.

4.3 *Alternative Probes of Dynamics for FDH and Other Enzymes*

In studying the potential functional roles of enzyme dynamics, it is critical to be able to observe the dynamics of complexes that mimic the TRS. In that sense, azide anion was an ideal probe for the dynamics of FDH because it just happens to be a transition-state-analog inhibitor. To study other FDH complexes we need to find an alternative chromophore that will allow us to replace the azide with other inhibitors to test our hypothesis further. An alternative probe would have the added benefit of exploring the dynamics of a different part of the protein active site to test whether the rigid structure that we have observed for the ternary complex spans the active site of the protein or if this effect is localized to a few residues. In addition, because azide will not mimic the TRS in other systems and only binds a limited number of enzymes, other probes would also offer access to the dynamics in other systems. To this end, we have prepared and characterized azido-labeled analogs of NAD^+, azido-NAD^+ [103, 104], in which the amide group on the nicotinamide ring has been replaced with an azide, and picolyl azide adenine dinucleotide ($PAAD^+$) in which the amide group is replaced by an azide that is separated from the pyridinium ring by a methylene spacer [105]. Azido-NAD^+ is attractive because it is a smaller perturbation than is $PAAD^+$, but $PAAD^+$ has a significantly stronger transition moment. We have characterized the binding and reactivity of azido-NAD^+ with several enzymes. It is an inhibitor for malate dehydrogenase but is an active substrate for FDH and for glucose dehydrogenase. For all of these enzymes, the binding affinity of azido-NAD^+ is comparable to that for NAD^+. For $PAAD^+$ we have only studied the interactions with FDH. $PAAD^+$ is an inhibitor of FDH with a binding affinity similar to that of NAD^+. Although azido-NAD^+ is such a weak chromophore that measurements of enzyme-bound azido-NAD^+ are extremely challenging, we have reported experiments on the binary complex of $PAAD^+$ with FDH suggesting that it will make an excellent probe of enzyme dynamics. As would be expected based on our earlier measurements of FDH with azide, we see a substantial static component in the FFCF decay of the binary complex of $PAAD^+$ with FDH reflecting the fact that this binary complex is not a good mimic of the TRS. Further experiments on the ternary complex of FDH with $PAAD^+$ and azide are underway to test (1) whether the extreme rigidity probed by the azide for the ternary complex with NAD^+ is present throughout the active site or is localized to the residues in the immediate vicinity of the H-transfer and (2) whether the dynamics of other ternary complexes of FDH (i.e., not transition-state-analog complexes) exhibit a static component to the FFCF indicating that they are not transition state structures consistent with our hypothesis that rigidity is a dynamic signature of the transition state complex.

5 Future Directions

5.1 *Methods*

5.1.1 New Chromophores

Among the most significant challenges to further applications of 2D IR spectroscopy to the study of enzyme dynamics, the greatest is finding suitable chromophores, following the four criteria specified in Sect. 1.3. Heme–CO has been a very productive label because it meets all of the criteria for 2D IR perfectly: a strong transition moment, a transition in the open water window, a long population lifetime, and a binding mechanism that makes it suitable for a range of proteins. None of the other probes that have been used to date have all of these features. Although azide has been particularly useful in the study of FDH, it is fortuitous that azide happens to be a transition-state-analog inhibitor for this enzyme. $PAAD^+$ represents a significant step towards having a generally applicable probe of NAD^+-dependent enzymes. This same approach of labeling a cofactor has also been used recently by Tucker et al. who have synthesized and characterized 2′-azido-2′-deoxyuridine [106], and it seems likely that this approach will gain more widespread use in 2D IR studies. Unfortunately, the azido antisymmetric stretching vibration has a relatively short population lifetime, limiting the range of time scales that are accessible by 2D IR. This basic strategy of labeling a cofactor could, however, be applied with different chromophores. Among the most promising is the thiocyanate label. This label can be incorporated in a cofactor as we have done with azide or it can be incorporated in the protein sequence by introducing a cysteine mutation and converting the thiol to thiocyanate, an approach which has been used to study measure infrared absorption spectra and Stark shifts in a variety of proteins [107–112]. Unfortunately, although the thiocyanate CN-stretching lifetime is quite long, the transition moment is quite weak. The molar absorptivity of the CN stretch of alkyl thiocyanates, which is proportional to the square of the transition dipole, is an order of magnitude weaker than that for the azide anion. Since the signal of a 2D IR experiment scales as the transition dipole to the fourth power, we would expect thiocyanates to give a signal that is 100-fold weaker than azide. At the few mM concentrations that are accessible with most proteins, this signal strength poses a significant challenge. Nevertheless, thiocyanates remain an attractive probe vibration for biomolecular dynamics. Other possible probe vibrations include nitriles, which were used in the HIV-1 reverse transcriptase experiments from the Hochstrasser group and C–D stretching vibrations. Nitriles, like azides, suffer from a short population lifetime and also have a weak transition moment, suggesting that they may not be the most suitable probes for 2D IR experiments. C–D stretching vibrations are very attractive because isotope labeling a C–H is a very minor perturbation. The C–D population lifetime of alkyl side chains are likely to be somewhat longer than that for azides or nitriles, but the transition moments is much weaker even than nitriles or thiocyanates placing C–D stretching vibrations well outside of what is currently accessible by 2D IR methods.

5.1.2 New Cofactors

New cofactors will enable the examination of enzymes that do not utilize nicotinamide cofactors. Until we are able to study C–D stretches, a triple bond like those mentioned above will be required. An example of appropriate cofactor targets would include derivatives of folic acid that are used by methylating enzymes (e.g., dihydrofolate, tetrahydrofolate, methylene-tetrahydrofolate, folinic acid, etc.). The folic acid backbone includes a *p*-aminobenzoic acid moiety, which can be substituted at the meta or ortho positions by cyano, azido, nitrile, thiocyanates, or other IR chromophores. Another example would be derivatives of flavin (e.g., FAD or FMN) modified by one of these IR chromophores at the eighth carbon. This carbon is on the phenolic ring that is conjugated but not part of the redox system, and is commonly modified by various moieties of similar size to the chromophores mentioned above. In accordance with the use of 2′-azido-2′-deoxyuridine [106] for studies of RNA related enzymes, one can imagine development of 3′-azido-2′,3′-dideoxythymidine (3′-azido-dTMP) to study DNA related enzymes. Furthermore, labeling of dTMP with IR chromophores at the C5 position is also synthetically feasible and those may also serve as probes for 2D IR studies in some enzymes related to DNA and its biosynthesis. Naturally, some of these modified cofactors will not serve as alternative substrates and might not even bind to a particular enzyme. Thus a search for a derivative with appropriate binding characteristics will have to precede any IR study to ensure that the measured dynamics are relevant for the enzyme and its catalyzed reaction.

5.2 *Future Applications*

5.2.1 Mutantion Studies

In order to examine the possible correlations between the dynamics measured for FDH and the influence of the DAD on the chemistry as examined by TDKIEs, we will examine serial hydrophobic mutations of residues holding the nicotinamide (the H-acceptor in this case) close to the H-donor (i.e., V123 and I175 in Candida FDH; see Fig. 9). These studies will follow the pattern of the kinetic studies of I14 mutants of DHFR (Sect. 2.1), but will add 2D IR spectroscopy of the same mutants in order to probe for correlation between the static component of the FFCF or one of the other parameters and the temperature dependence of the KIE. Such correlation may afford the first direct evidence for the relationship between the active-site dynamics and the hydride-transfer kinetics.

Once an NADPH derivative with an appropriate IR chromophore (Sect. 1.3) is prepared, similar studies of DHFR including both active site and remote mutations as discussed in Sect. 2.1 will be the subject of spectroscopic investigations to probe for similar correlations as discussed for FDH.

5.2.2 Other Nicotinamide Dependent Enzymes

The derivatives presented in Sect. 5.1 will be screened for either activity or inhibition of a broad range of nicotinamide-dependent enzymes. Successful candidates will undergo investigations similar to those proposed above for FDH and DHFR, which will assist in assessing how general are the correlations identified for these two systems. This study may lead to a more general understanding of the relationship

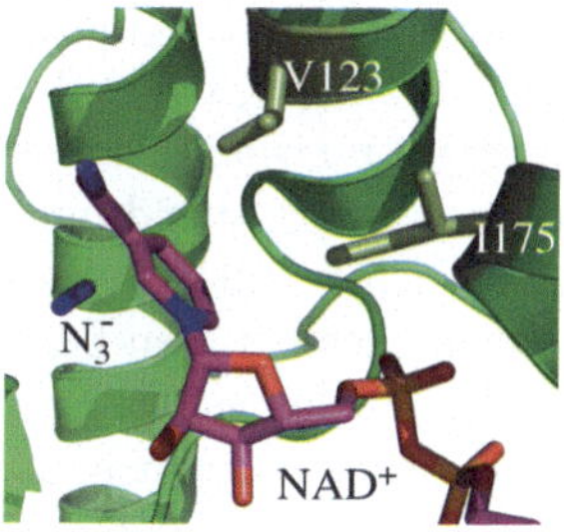

Fig. 9 Active site of FDH with N_3^- and NAD^+, with V123 and I175 highlighted

between fast enzyme dynamics and the catalyzed chemistry, i.e., whether there is a general role for dynamics in all/many systems or whether different systems can be divided into different groups with different roles for the enzyme motions.

6 Summary

As discussed above and in several of the other chapters in this issue, protein motions on different time scales play a critical role in enzyme function. So far, most studies on the femtosecond to picosecond time scale have addressed the effect of protein denaturation, ligation to substrates and inhibitors, mutations, modification of S–S bonds, and His tag labeling on active site dynamics. These studies are described in Sect. 3 and in general seem to indicate that proteins rigidify upon binding of tight-binding ligands, that this and other perturbations can alter dynamics across the protein, even tens of Angstroms away from the perturbation, and that those effects on the femtosecond to picosecond time scale are consistent with those reported for nanosecond to microsecond time scale motions as detected by NMR relaxation studies. Since the physical and molecular character of the fast dynamics is different from that of the slower motions, these findings are neither intuitive nor trivial. The ability of protein modifications to alter dynamics on the fast time scale even if it seems not to lead to any alteration of average structure (e.g., Sect. 3.2) provides experimental support for MD simulations suggesting such effects [42, 51]. These results may also explain the apparent transmission of altered dynamics across enzymes as suggested by kinetic studies of remote mutants [41].

Motions that are most likely to affect the activation of covalent bonds (C–H in this case) are likely to be on the same time scale as that process, i.e., femtosecond–picosecond. Additionally, the chemical step has to be exposed, despite taking place on such a fast time scale relative to steps associated with ligand binding, protein motions that bring the reactants close together, product release, etc. To address the first issue, 2D IR spectroscopy was developed to enable examination of motions on the fs–ps time scale in transition state or ground state analogs of the reaction under study. To address the second issue, kinetic methods were developed that expose the nature of the C–H → C transfer catalyzed by several enzymes. These methods appear sensitive to the dynamics and orientation of the H-donor and H-acceptor. Recent studies of FDH demonstrate that this combination of spectroscopic and kinetic methods implicate well-oriented donor and acceptor for H-tunneling at the TRS (using a transition state analog and intrinsic KIEs). Both dynamic and kinetic data suggest very fast and rigid dynamics of the active site that assure an accurate orientation of the H-donor and acceptor for the reaction catalyzed by FDH. To probe correlations between the bond activation (studied kinetically) and its environmental dynamics (studied spectroscopically), rigorous site directed mutagenesis that alters the reaction's environment as described in Sect. 5, will be conducted and examined by both methods. Initially, FDH mutants will be studied using azide in complex with NAD^+, $PAAD^+$, or azido-NAD^+, as discussed in detail in Sects. 4 and 5. Then the dynamics of DHFR and its active site mutants will be studied using these alternative co-factors (but with 2′-phosphate, i.e., $PAADP^+$ and azido-$NADP^+$). A better understanding of the relationship between enzyme-catalyzed bond activation and the enzyme dynamics that affect that process is likely to impact our conception of the physical features by which enzymes activate covalent bonds, impact theoretical calculations and simulations that examine both processes, direct the search for inhibitors based on both the catalyzed chemistry and structure of the reactive species, and affect rational design of biomimetic catalysts.

References

1. Schowen RL (2009) The strengths and weaknesses of model reactions for the assessment of tunnelling in enzymic reactions. In: Allemann R, Scrutton N (eds) Quantum tunnelling in enzyme catalyzed reactions. Royal Society of Chemistry, London, pp 292–313 (Chap. 13)
2. Boekelheide N, Salomón-Ferrer R, Miller TF (2011) Proc Natl Acad Sci U S A 108:16159–16163
3. Hammes-Schiffer S (2006) Acc Chem Res 39:93–100
4. Cleland WW (2006) Enzyme mechanisms from isotope effects. In: Kohen A, Limbach HH (eds) Isotope effects in chemistry and biology. Taylor & Francis/CRC, Boca Raton, pp 915–930 (Chap. 37)
5. Cook PF (1991) Kinetic and regulatory mechanisms of enzymes from isotope effects. In: Cook PF (ed) Enzyme mechanism from isotope effects. CRC, Boca Raton, pp 203–230
6. Cook PF, Cleland WW (2007) Enzyme kinetics and mechanism. Taylor & Francis Group LLC, New York
7. Bahnson BJ, Park DH, Kim K, Plapp BV, Klinman JP (1993) Biochemistry 32:5503–5507
8. Northrop DB (1991) Intrinsic isotope effects in enzyme catalyzed reactions. In: Cook PF (ed) Enzyme mechanism from isotope effects. CRC, Boca Raton, pp 181–202
9. Sen A, Yahashiri A, Kohen A (2011) Biochemistry 50:6462–6468
10. Swain CG, Stivers EC, Reuwer JF, Schaad LJ (1958) J Am Chem Soc 80:5885–5893
11. Kohen A, Klinman JP (1998) Acc Chem Res 31:397–404
12. Kohen A, Klinman JP (1999) Chem Biol 6:R191–R198
13. Bell RP (1980) The tunnel effect in chemistry. Chapman & Hall, London
14. Basran J, Patel S, Sutcliffe MJ, Scrutton NS (2001) J Biol Chem 276:6234–6242
15. Limbach HH, Lopez JM, Kohen A (2006) Phil Trans R Soc B Biol Sci 361:1399–1415
16. Marcus RA, Sutin N (1985) Biochim Biophys Acta 811:265–322
17. Kohen A, Roston D, Stojković V, Wang Z (2011) Kinetic isotope effects in enzymes. In: Meyers RA (ed) Encyclopedia of analytical chemistry, vols. S1–S3. Wiley, Chichester, pp 77–99

18. Knapp MJ, Klinman JP (2002) Eur J Biochem 269:3113–3121
19. Nagel ZD, Klinman JP (2010) Chem Rev 110:PR41–PR67
20. Schwartz SD (2006) Vibrationally enhanced tunneling from the temperature dependence of KIE. In: Kohen A, Limbach HH (eds) Isotope effects in chemistry and biology. CRC, Boca Raton, pp 475–498 (Chap. 18)
21. Kuznetsov AM, Ulstrup J (1999) Can J Chem 77:1085–1096
22. Pudney CR, Johannissen LO, Sutcliffe MJ, Hay S, Scrutton NS (2010) J Am Chem Soc 132:11329–11335
23. Borgis DC, Lee SY, Hynes JT (1989) Chem Phys Lett 162:19–26
24. Sen A, Kohen A (2010) J Phys Org Chem 23:613–619
25. Bhabha G, Lee J, Ekiert DC, Gam J, Wilson IA, Dyson HJ, Benkovic SJ, Wright PE (2011) Science 332:234–238
26. Boehr DD, McElheny D, Dyson HJ, Wright PE (2006) Science 313:1638–1642
27. Adamczyk AJ, Cao J, Kamerlin SCL, Warshel A (2011) Proc Natl Acad Sci U S A 108:14115–14120
28. Pisliakov AV, Cao J, Kamerlin SCL, Warshel A (2009) Proc Natl Acad Sci U S A 106:17359–17364
29. Hamm P, Zanni MT (2011) Concepts and methods of 2D infrared spectroscopy. Cambridge University Press, New York
30. Kwak K, Park S, Finkelstein IJ, Fayer MD (2007) J Chem Phys 127:124503
31. Kwak K, Rosenfeld DE, Fayer MD (2008) J Chem Phys 128:204505
32. Liu H, Warshel A (2007) J Phys Chem B 111:7852–7861
33. Garcia-Viloca M, Truhlar DG, Gao JL (2003) J Mol Biol 327:549–560
34. Garcia-Viloca M, Truhlar DG, Gao J (2003) Biochemistry 42:13558–13575
35. Garcia-Viloca M, Gao J, Karplus M, Truhlar DG (2003) Science 303:186–195
36. Benkovic SJ, Hammes-Schiffer S (2003) Science 301:1196–1202
37. Hammes-Schiffer S, Benkovic SJ (2006) Annu Rev Biochem 75:519–541
38. Hammes-Schiffer S (2006) Kinetic isotope effects for proton-coupled electron transfer reactions. In: Kohen A, Limbach HH (eds) Isotope effects in chemistry and biology. CRC, Boca Raton, pp 499–520
39. Miller GP, Benkovic SJ (1998) Chem Biol 5:R105–R113
40. McElheny D, Schnell JR, Lansing JC, Dyson HJ, Wright PE (2005) Proc Natl Acad Sci U S A 102:5032–5037
41. Wang L, Goodey NM, Benkovic SJ, Kohen A (2006) Proc Natl Acad Sci U S A 103:15753–15758
42. Wong KF, Selzer T, Benkovic SJ, Hammes-Schiffer S (2005) Proc Natl Acad Sci U S A 102:6807–6812
43. Farnum MF, Magde D, Howell EE, Hirai JT, Warren MS, Grimsley JK, Kraut J (1991) Biochemistry 30:11567–11579
44. Epstein DM, Benkovic SJ, Wright PE (1995) Biochemistry 34:11037–11048
45. Sawaya MR, Kraut J (1997) Biochemistry 36:586–603
46. Wang L, Tharp S, Selzer T, Benkovic SJ, Kohen A (2006) Biochemistry 45:1383–1392
47. Wang L, Goodey NM, Benkovic SJ, Kohen A (2006) Phil Trans R Soc B Biol Sci 361:1307–1315
48. Sikorski RS, Wang L, Markham KA, Rajagopalan PTR, Benkovic SJ, Kohen A (2004) J Am Chem Soc 126:4778–4779
49. Stojković V, Perissinotti LL, Lee J, Benkovic SJ, Kohen A (2012) J Am Chem Soc 134:1738–1745
50. Creighton TE (1984) Proteins. Structure and molecular principles. Freeman, New York
51. Rod TH, Brooks CL (2003) J Am Chem Soc 125:8718–8719
52. Bruice TW, Santi DV (1991) Isotope effects in reactions catalyzed by thymidylate synthase. In: Cook PF (ed) Enzyme mechanism from isotope effects. CRC, Boca Raton, pp 457–479
53. Carreras CW, Santi DV (1995) Annu Rev Biochem 64:721–762

54. Stroud RM, Finer-Moore JS (2003) Biochemistry 42:239–247, and references cited therein
55. Finer-Moore JS, Santi DV, Stroud RM (2003) Biochemistry 42:248–256, and references cited therein
56. Agrawal N, Hong B, Mihai C, Kohen A (2004) Biochemistry 43:1998–2006
57. Hong B, Haddad M, Maley F, Jensen JH, Kohen A (2006) J Am Chem Soc 128:5636–5637
58. Hong B, Maley F, Kohen A (2007) Biochemistry 46:14188–14197
59. Kanaan N, Martí M, Moliner V, Kohen A (2006) Biochemistry 46:3704–3713
60. Kanaan N, Martí M, Moliner V, Kohen A (2009) J Phys Chem A 113:2176–2182
61. Kanaan N, Martí M, Moliner V, Kohen A (2007) Biochemistry 46:3704–3713
62. Kanaan N, Ferrer S, Martí S, Garcia-Viloca M, Kohen A, Moliner V (2011) J Am Chem Soc 133:6692–6702
63. Wang Z, Kohen A (2010) J Am Chem Soc 132:9820–9825
64. Spencer HT, Villafranca JE, Appleman JR (1997) Biochemistry 36:4212–4222
65. Lomax S, Greenberg MI, Robert G (1967) J Biol Chem 242:109–113
66. Masgrau L, Roujeinikova A, Johannissen LO, Hothi P, Basran J, Ranaghan KE, Mulholland AJ, Sutcliffe MJ, Scrutton NS, Leys D (2006) Science 312:237–241
67. Hay S, Pudney CR, Sutcliffe MJ, Scrutton NS (2008) Angew Chem Int Ed 47:537–540
68. Hay S, Sutcliffe MJ, Scrutton NS (2007) Proc Natl Acad Sci USA 104:507–512
69. Kohen A, Cannio R, Bartolucci S, Klinman JP (1999) Nature 399:496–499
70. Kiefer PM, Hynes JT (2003) J Phys Chem A 107:9022–9039
71. Nagel ZD, Meadows CW, Dong M, Bahnson BJ, Klinman JP (2012) Biochemistry 51:4147–4156
72. Kwart H (1982) Acc Chem Res 15:401–408
73. Braun J, Schwesinger R, Williams PG, Morimoto H, Wemmer DE, Limbach HH (1996) J Am Chem Soc 118:11101–11110
74. Thielges MC, Chung JK, Fayer MD (2011) J Am Chem Soc 133:3995–4004
75. Thielges MC, Chung JK, Axup JY, Fayer MD (2011) Biochemistry 50:5799–5805
76. Chung JK, Thielges MC, Bowman SEJ, Bren KL, Fayer MD (2011) J Am Chem Soc 133:6681–6691
77. Bagchi S, Nebgen BT, Loring RF, Fayer MD (2010) J Am Chem Soc 132:18367–18376
78. Kim S, Chung JK, Kwak K, Bowman SEJ, Bren KL, Bagchi B, Fayer MD (2008) J Phys Chem B 112:10054–10063
79. Ishikawa H, Kwak K, Chung JK, Kim S, Fayer MD (2008) Proc Natl Acad Sci U S A 105:8619–8624
80. Ishikawa H, Kim S, Kwak K, Wakasugi K, Fayer MD (2007) Proc Natl Acad Sci U S A 104:19309–19314
81. Ishikawa H, Finkelstein IJ, Kim S, Kwak K, Chung JK, Wakasugi K, Massari AM, Fayer MD (2007) Proc Natl Acad Sci U S A 104:16116–16121
82. Finkelstein IJ, Massari AM, Fayer MD (2007) Biophys J 92:3652–3662
83. Finkelstein IJ, Ishikawa H, Kim S, Massari AM, Fayer MD (2007) Proc Natl Acad Sci U S A 104:2637–2642
84. Massari AM, Finkelstein IJ, McClain BL, Goj A, Wen X, Bren KL, Loring RF, Fayer MD (2005) J Am Chem Soc 127:14279–14289
85. Finkelstein IJ, Goj A, McClain BL, Massari AM, Merchant KA, Loring RF, Fayer MD (2005) J Phys Chem B 109:16959–16966
86. Fang C, Bauman JD, Das K, Remorino A, Arnold E, Hochstrasser RM (2008) Proc Natl Acad Sci U S A 105:1472–1477
87. Lim MH, Hamm P, Hochstrasser RM (1998) Proc Natl Acad Sci U S A 95:15315–15320
88. Thielges MC, Axup JY, Wong D, Lee HS, Chung JK, Schultz PG, Fayer MD (2011) J Phys Chem B 115:11294–11304
89. Bandaria JN, Dutta S, Nydegger MW, Rock W, Kohen A, Cheatum CM (2010) Proc Natl Acad Sci U S A 107:17974–17979
90. Bandaria JN, Dutta S, Hill SE, Kohen A, Cheatum CM (2008) J Am Chem Soc 130:22–23

91. Hill SE, Bandaria J, Fox M, Vanderaugh E, Kohen A, Cheatum CM (2009) J Phys Chem B 113:11505–11510
92. Lindquist BA, Furse KE, Corcelli SA (2009) Phys Chem Chem Phys 11:8119–8132
93. Kuo CH, Hochstrasser RM (2007) Chem Phys 341:21–28
94. Hamm P, Lim M, Hochstrasser RM (1998) Phys Rev Lett 81:5326–5329
95. Fecko CJ, Loparo JJ, Roberts ST, Tokmakoff A (2005) J Chem Phys 122:054506
96. Eaves JD, Tokmakoff A, Geissler PL (2005) J Phys Chem A 109:9424–9436
97. Asbury JB, Steinel T, Kwak K, Corcelli SA, Lawrence CP, Skinner JL, Fayer MD (2004) J Chem Phys 121:12431–12446
98. Fecko CJ, Eaves JD, Loparo JJ, Tokmakoff A, Geissler PL (2003) Science 301:1698–1702
99. Venkitakrishnan RP, Zaborowski E, McElheny D, Benkovic SJ, Dyson HJ, Wright PE (2004) Biochemistry 43:16046–16055
100. Osborne MJ, Schnell J, Benkovic SJ, Dyson HJ, Wright PE (2001) Biochemistry 40:9846–9859
101. Bandaria J, Cheatum C, Kohen A (2009) J Am Chem Soc 131:10151–10155
102. Lamzin VS, Dauter Z, Popov VO, Harutyunyan EH, Wilson KS (1994) J Mol Biol 236:759–785
103. Dutta S, Rock W, Cook RJ, Kohen A, Cheatum CM (2011) J Chem Phys 135:055106 (6)
104. Dutta S, Cook RJ, Houtman JCD, Kohen A, Cheatum CM (2010) Anal Biochem 407:241–246
105. Dutta S, Li Y-L, Rock W, Houtman JCD, Kohen A, Cheatum CM (2012) J Phys Chem B 116:542–548
106. Tucker MJ, Gai XS, Fenlon EE, Brewer SH, Hochstrasser RM (2011) Phys Chem Chem Phys 13:2237–2241
107. Fafarman AT, Boxer SG (2010) J Phys Chem B 114:13536–13544
108. Fafarman AT, Webb LJ, Chuang JI, Boxer SG (2006) J Am Chem Soc 128:13356–13357
109. Hu W, Webb LJ (2011) J Phys Chem Lett 2:1925–1930
110. Stafford AJ, Ensign DL, Webb LJ (2010) J Phys Chem B 114:15331–15344
111. McMahon HA, Alfieri KN, Clark CAA, Londergan CH (2010) J Phys Chem Lett 1:850–855
112. Maienschein-Cline MG, Londergan CH (2007) J Phys Chem A 111:10020–10025

Top Curr Chem (2013) 337: 41–68
DOI: 10.1007/128_2012_411

Published online: 28 March 2013

Protein Conformational Disorder and Enzyme Catalysis

Cindy Schulenburg and Donald Hilvert

Abstract Though lacking a well-defined three-dimensional structure, intrinsically unstructured proteins are ubiquitous in nature. These molecules play crucial roles in many cellular processes, especially signaling and regulation. Surprisingly, even enzyme catalysis can tolerate substantial disorder. This observation contravenes conventional wisdom but is relevant to an understanding of how protein dynamics modulates enzyme function. This chapter reviews properties and characteristics of disordered proteins, emphasizing examples of enzymes that lack defined structures, and considers implications of structural disorder for catalytic efficiency and evolution.

Keywords Dynamics · Evolution · Intrinsic disorder · Protein function · Protein structure

Contents

C. Schulenburg and D. Hilvert (✉)
Laboratory of Organic Chemistry, ETH Zürich, 8093 Zürich, Switzerland
e-mail: donald.hilvert@org.chem.ethz.ch

1 Introduction

It is axiomatic in biochemistry that proteins must adopt a stable, well-defined tertiary fold to be wholly functional. Given this, sequence–structure–function analyses have generally concentrated on fully structured proteins. In addition to a stable fold, however, proteins require certain flexibility. Because fluctuating interactions of defined amino acid residues or conformational movements of entire loops or domains are crucial to many biological activities, interest is mounting in proteins that present increased conformational flexibility and even lack stable tertiary structure.

Amino acid sequences dictate protein conformation, but not every amino acid sequence is able to produce a stable tertiary structure and not all proteins within a cell are well structured. Proteins that cannot adopt a stable tertiary fold are called natively unfolded, intrinsically unfolded, or intrinsically disordered. Such molecules are not typically fully unstructured. Local residual secondary structures may be present but insufficient to stabilize long-range tertiary interactions under physiological conditions. High intrinsic flexibility results in dynamic ensembles of rapidly interconverting conformational states of comparable energy [1]. Within these ensembles, individual polypeptide chains may adopt collapsed (molten globule) or extended, random coil-like (pre-molten globule) conformations (Fig. 1). Other proteins may contain partially disordered regions. When positioned at protein termini, flexible segments can serve as signal sequences. When located within multi-domain proteins they may function as dynamic linkers connecting individual domains.

Intrinsically disordered proteins (IDPs) and proteins with large intrinsically disordered regions are found surprisingly frequently in all domains of life. Interestingly, the proportion of IDPs and proteins containing disordered regions increases with organismal complexity [3]. Algorithms based on the biased amino acid compositions and specific biochemical properties of IDPs predict that approximately one-third of

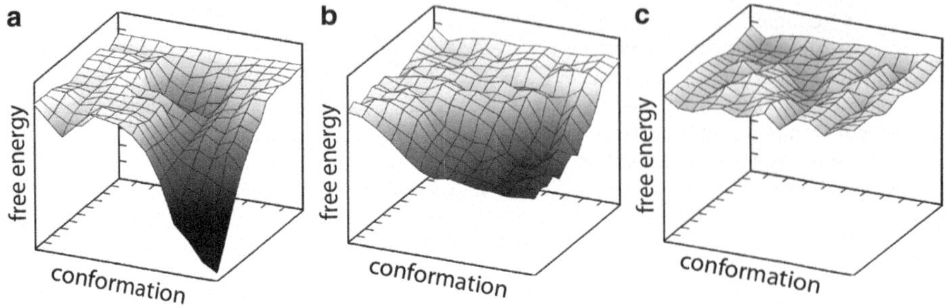

Fig. 1 Energy landscapes for increasingly disordered proteins. (**a**) The native conformation of a well-folded protein has a well-defined minimum energy. (**b**) A molten globular protein exhibits multiple, partially folded conformations. (**c**) An intrinsically unstructured protein lacks a deep minimum and consists of many different interconverting species of comparable free energy. The figure is partially adapted from [2]

all eukaryotic proteins are partly or completely disordered [4, 5]. In mammals, one quarter is fully disordered and half contains extended disordered regions [6]. Disordered proteins play important roles in many fundamental biochemical processes, ranging from transcription and translation to signal transduction and regulation [7]. Even enzymes, the biological catalysts responsible for accelerating nearly all metabolic reactions in the cell, can exhibit significant structural disorder.

In this review we examine the properties of intrinsically disordered proteins with special emphasis on enzyme catalysis. The implications of intrinsic disorder for enzymatic efficiency and evolution are considered.

2 Intrinsically Disordered Proteins

Conformational flexibility is an inherent property of all polypeptide sequences, whether folded or non-folded. The diversity of motions observed in proteins extends from local fluctuations of amino acid side chains and loops on the picosecond to nanosecond timescale to domain movements and complete rearrangements of entire protein folds on the microsecond to second timescale [8]. Such jostling may give rise to altered protein conformations at any time, but most states represent modest structural excursions around the native protein fold [9]. Because motions on different timescales are coupled, local fluctuations in the picosecond to nanosecond time scale can influence larger structural changes [10]. Intrinsic disorder takes these normal dynamic attributes of proteins to extremes.

Both protein folding and protein non-folding are encoded at the level of amino acid sequence [11]. Disordered proteins typically exhibit relatively low sequence complexity compared to well-ordered proteins [12]. They contain few hydrophobic amino acid residues [1], which are usually buried in ordered protein cores and favor formation of compact structures in hydrophilic environments. Many polar and charged amino acid residues lead to high net polypeptide chain charge. Owing to electrostatic repulsion, expanded rather than compact structures result [13–15]. In fact, plotting mean protein hydrophobicity as a function of mean net charge provides a useful means of identifying IDPs at the proteome level [16]. Methionine, which has a flexible aliphatic side chain, and proline, which disrupts secondary structure, are overrepresented in IDPs. Repeated sequences, such as polyalanine or polyglycine stretches, are also prevalent [1].

Given low structural content and expanded structure, disordered proteins have distinctive biophysical properties. For instance, they typically elute from gel filtration columns as broad peaks, unfold non-cooperatively [17–19], and evince greater susceptibility to proteolytic attack than ordered proteins [20]. As hydrophobic residues in unstructured proteins are largely accessible to solvent, protein disorder can often be detected with dyes like 1-anilino-naphthalene-8-sulphonate (ANS) or thioflavin T that bind to exposed hydrophobic patches [21]. Although protein crystallization imparts little useful structural information about unstructured regions, small angle X-ray scattering and NMR spectroscopy can be profitably exploited to characterize IDPs [22–26]. Their dynamic properties typically cause peak broadening and/or low signal dispersion in 1D and 2D NMR spectra [18, 27–29]. Fast hydrogen/deuterium exchange, which can be detected by either NMR spectroscopy or mass spectrometry, is another consequence of intrinsic disorder [30].

3 Target Recognition and Binding

Although the notion that three-dimensional structure is required for protein function has historically dominated the thinking of chemists and biologists, it is now clear that proteins lacking a defined tertiary structure play a surprising diverse set of biochemical roles. As noted above, many molecules involved in signaling and regulation of the cell cycle are intrinsically disordered [31]. It has been estimated that ~75% of all mammalian proteins involved in cell signaling and regulation contain large unstructured regions or are fully disordered [6]. Transcription and translation are additional key biological processes in which global or partial disorder figures [32, 33]. As a consequence, disorder often figures in disease [34]. Plants and insects utilize such proteins as protection against dehydration and for regulation of intracellular salt concentrations [35]. IDPs can also serve as degradation tags [33] or as chaperones [36, 37].

Typically, IDP activity is expressed through interactions with a specific partner, such as DNA, RNA, or another protein. Upon binding to a target, most IDPs undergo significant conformational change, typically leading to a more ordered state [38]. In the extreme case, these disorder-to-order transitions result in conversion of a random coil into a well-defined three-dimensional structure. Both the binding and folding steps are entropically unfavorable. As unbound IDPs are rarely completely unfolded [39], however, the initial conformational space is not as great as that expected for a fully unstructured polypeptide chain. Moreover, large regions of an IDP may remain highly flexible even when bound. As in more conventional protein folding, the entropic costs of conformational reorganization in these cases are (partially) offset by entropic gains associated with the release of ordered water molecules and by favorable enthalpic interactions within the folded structure. In addition, target-dependent disorder-to-order transitions benefit from intermolecular interactions between the IDP and its binding partner.

The precise mechanism of IDP binding is still debated. A conformational selection model (Fig. 2a) posits rapid interconversion of many different states for an IDP, only one or a few of which are correctly configured for target recognition. Formation of a productive complex shifts the equilibrium toward the folded conformation, which becomes increasingly populated over time [40]. As only two populations – bound and unbound – are kinetically relevant, folding is thought to follow a simple two-state mechanism [41]. A second theory postulates that folding is coupled to target binding (Fig. 2b). In this scenario, binding can take place at any number of initiation sites along the IDP polypeptide chain and induces folding [42]. Parallel folding pathways and multiple structures may result [41].

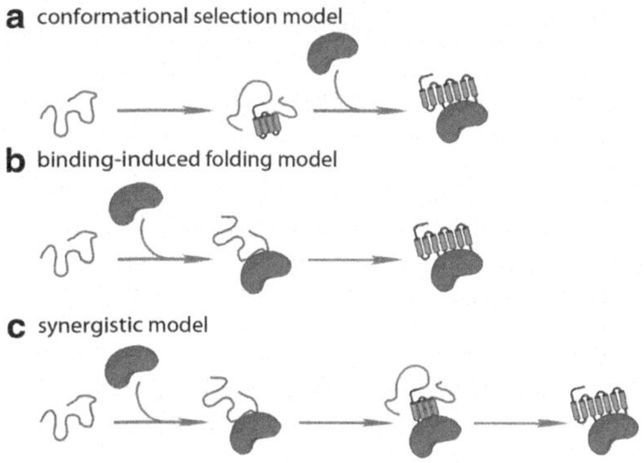

Fig. 2 Models for target-dependent folding of intrinsically disordered proteins. (**a**) In the conformational selection model the target molecule binds to a specific subspecies of the intrinsically disordered protein. (**b**) In the binding-induced folding model the intrinsically disordered protein folds upon binding to the target molecule. (**c**) The synergistic model combines aspects of the conformational selection and binding-induced folding models

By analyzing the folding behavior of an IDP in the presence of its binding partner it is sometimes possible to distinguish between the different binding models. Examples of each have been described in the literature [24, 43–46]. In many cases, however, the mechanisms cannot be differentiated. Sometimes, multiple mechanisms may be operative. To account for this observation, a synergistic model for IDP folding (Fig. 2c) was proposed that combines the two pathways [47]. The relative contribution of the individual mechanisms depends on case-specific factors such as binding rates, IDP concentration, protein plasticity, and the specific nature of the disorder-to-order transition. Some IDPs are thought to exploit different binding modes in different environments or in response to different binding partners. For example, the C-terminal segment of tumor suppressor protein p53 adopts four types of secondary structure when complexed with different binding partners [48]. The conformational states and binding modes of this protein are further modulated by posttranslational phosphorylation, methylation, and acylation [49, 50].

Lack of stable tertiary structure and high inherent flexibility confer multiple advantages on IDPs compared to structured proteins with respect to molecular recognition. Almost all biological functions of these proteins depend on an appropriate balance of specificity and affinity for different targets [51]. The increased interaction surface and conformational fluctuations of IDPs can be exploited to maximize the rate of substrate binding and product release. The larger capture radius of an unstructured protein allows it to bind weakly and nonspecifically to its target at large distances – as in "fly-casting" – and fold as it approaches the actual binding site [52]. Consequently, IDPs require fewer encounters than structured proteins to form complexes, accelerating the binding process [53]. Computer simulations suggest that linking binding and folding significantly reduces the free energy barrier for binding [47]. Nonetheless, increased flexibility of the polypeptide chain in the bound state also yields high k_{off} rates relative to fully folded proteins. The resulting rapid turnover rates are likely to be kinetically advantageous for the dynamic biological processes involving IDPs.

When coupled, high binding specificity and moderate binding affinity enable rapid and tunable regulation of the cell cycle and other biological processes in response to changing environmental conditions [54]. The extended conformations of IDPs enhance binding diversity. Indeed, IDPs often co-operate with multiple binding partners, forming hubs, or nodes, of large protein–protein interaction networks [55, 56]. Such hubs are so essential to key biochemical functions that deletion of the IDP can result in organism death. Alternatively, interaction networks can be established by binding of multiple IDPs to a single folded partner [57].

4 Intrinsically Disordered Enzymes

Catalysis poses a particularly demanding molecular recognition challenge. Biological enzymes make life possible by accelerating all metabolic reactions in the cell by enormous factors. Pauling postulated that protein catalysts lower the activation barrier of the reaction they promote by binding the transition state more tightly than the substrate(s) [58]. In this view, enzymes utilize a defined structure and preorganized sets of functional groups to provide maximum shape and chemical complementarity to this transient, high-energy species, selectively stabilizing it.

Recognition that proteins are not rigid led Koshland to modify this model [59]. He proposed that substrate not only binds to the preformed enzyme active site but also brings about structural changes that orient the catalytic residues for productive reaction. Subsequent biochemical and crystallographic studies of many enzymes support this induced fit hypothesis [60]. In most textbooks the induced fit model is conventionally depicted as a conversion of one tight conformational ensemble (free enzyme) to another distinct ensemble (bound enzyme) through local substrate-mediated structural rearrangements. It is now evident, however, that enzymes, like other proteins, exhibit an extensive array of motions over many timescales [61]. In extreme cases they even evince significant structural disorder.

Several naturally occurring enzymes are either fully disordered or include large disordered regions (Table 1). RNase E, an endoribonuclease found in the *Escherichia coli* RNA degradosome, consists of an evolutionarily conserved, well-structured N-terminal region [65], yet much of its C-terminal region is intrinsically disordered judging from sequence analysis, CD spectroscopy, and X-ray studies [66]. While its catalytic center is located in an ordered N-terminal region, its disordered C-terminus is essential for mediating interactions with the RNA substrate and other binding partners in the RNA degradosome [67].

The hepatitis C virus NS3 protease (HCVP) is a partially unstructured Zn^{2+} dependent serine-protease [19]. In the presence of Zn^{2+} ions it cleaves viral precursor proteins in non-structured regions. While not directly involved in catalysis, the Zn^{2+} ion induces large conformational changes that yield stable, catalytically active enzyme conformations [68]. UreG, a GTPase that functions as a

Table 1 Representative examples of intrinsically disordered enzymes

Enzyme	Organism	Metal ion	Structural disorder	Catalytic activity	Reference
RNase E (nuclease)	*E. coli*	–	Partially disordered	$k_{cat} = 1.4\ s^{-1}$	[62]
HCVP (protease)	Hepatitis C virus	Zn^{2+}	Partially disordered	$k_{cat} = 2.7 \times 10^{-2}\ s^{-1}$	[19, 63]
UreG (GTPase)	*Bacillus pasteurii*	Zn^{2+}, Ni^{2+}	Largely disordered	$k_{obs} = 6.7 \times 10^{-4}\ s^{-1}$ [a]	[27]
UreG (GTPase)	*Mycobacterium tuberculosis*	Zn^{2+}, Ni^{2+}	Largely disordered	$k_{obs} = 1.7 \times 10^{-4}\ s^{-1}$ [a]	[64]
Anhydrin (endonuclease)	*A. avenae*	Mg^{2+}, Mn^{2+}	Largely disordered	$k_{obs} = 5.5 \times 10^{-4}\ s^{-1}$ [a]	[35]
TPPP/p25 (GTPase)	Human	Mg^{2+}	Largely disordered	$k_{obs} = 2.3 \times 10^{-4}\ s^{-1}$ [a]	[28]

[a]It is unclear whether these enzymes were assayed at saturating or subsaturating substrate concentrations, hence only observed rate constants are available

chaperone in nickel trafficking and urease activation, is another example [64, 69]. It exhibits rapidly interconverting conformational sub-states [17]. Transitions among these disordered ensembles occur non-cooperatively but reversibly. Binding of Zn^{2+} ions induces a conformational change and protein stabilization, but has no effect on GTPase activity [27]. Like many other intrinsically disordered proteins [57, 70], UreG appears to function as a scaffold protein, coordinating the binding of multiple partners.

A third catalytically active IDP, anhydrin, is found in the nucleus of the anhydrobiotic nematode *Aphelenchus avenae* [35]. Anhydrin discharges two entirely independent functions. In addition to acting as a chaperone to reduce protein aggregation, it is an endonuclease that acts on supercoiled, linear, and chromatin DNA. Though approximately ten times less active than T7 endonuclease I, its specific activity is sufficiently high for physiological function. While Ca^{2+} ions do not influence its catalytic activity, Mg^{2+} and Mn^{2+} ions enhance it. Even in its DNA bound state, anhydrin remains largely unstructured [35].

The tubulin polymerization promoting protein/p25 (TPPP/p25) is an unstructured brain-specific protein that induces aberrant tubulin assemblies in vitro [71]. Intracellular TPPP/p25 levels influence cell differentiation and proliferation and might be implicated in Parkinson's disease and other nervous system pathologies [72]. The TPPP/p25 amino acid sequence exhibits characteristics typical of IDPs [72]; CD measurements attest to extensive random-coil structure [73], whereas 1D and 2D NMR spectra indicate multiple protein conformations and increased protein flexibility [28, 74]. Extended disordered segments are located primarily at the N- and C-termini, but the mid-region also exhibits high flexibility. Sequence alignments have identified multiple potential GTP binding sites. The micromolar GTP binding affinity increases in the presence of Mg^{2+} ions crucial for TPPP/p25 GTPase activity. GTP hydrolysis by TPPP/p25 is comparable in rate to the intrinsic GTPase activity of other small G proteins [28].

Although the number of naturally disordered enzymes is still small, our perspective is strongly biased by the dominance of X-ray crystallography as a structural tool. Others may well come to light in the future through intensified NMR spectroscopic study of proteins in solution.

5 Inducibly Disordered Enzymes

Protein disorder is inducible in vitro under conditions such as high temperature, extreme pH, and distinct denaturant or salt concentrations [75]. In these circumstances many proteins lose their defined tertiary structure and biochemical function. On occasion, however, a modicum of activity can be retained or recovered by a disorder-to-order transition induced by substrate or another ligand (Table 2).

Disordered states are often transiently populated during protein folding [81]. For example, molten globule conformations are obligatory intermediates in the folding pathways of many proteins [82, 83]. Generally speaking, these highly dynamic protein states do not exhibit catalytic activity. In the case of *Sulfolobus solfataricus* acylphosphatase, however, an enzymatically active non-native folding intermediate has been identified [76]. Although its active site is structurally heterogeneous, the folding intermediate exhibits 80% of the catalytic activity of the fully folded enzyme. Control experiments showed that enzymatic activity does not derive from the fraction of native protein in the sample, but substrate-induced organization of the active site could not be ruled out. NMR studies on a homologous *Bacillus subtilis* acylphosphatase indicated a structurally disordered active site in the absence of substrate that became more ordered upon substrate binding [84]. Conformational heterogeneity was particularly pronounced at acidic pH, where the enzyme exhibits optimal activity, possibly accounting for the relatively broad specificity of the enzyme. Phosphate was found to alter the distribution of states present in solution by selectively binding to the active conformation.

Human topoisomerase I catalyzes topological changes in DNA by a two-step mechanism involving a covalent enzyme–DNA intermediate. The highly conserved C-terminal domain contains the active site nucleophile Tyr723, but lacks enzymatic activity on its own. Catalysis arises only in the presence of the core domain, which contains additional catalytically relevant residues. In solution, the isolated C-terminal domain adopts a largely α-helical molten-globule-like state whose protein surface is a virtual patchwork of hydrophobic elements [77]. Association with the core domain affords a complex that still exhibits some structural fluctuations. Within a narrow salt-concentration range, and despite 20-fold lower DNA affinity, this reconstituted complex maintains topoisomerase activity [78]. The conformational flexibility of the C-terminal domain may be biologically important for accommodating the large conformational changes that occur during the catalytic reaction [77].

Point mutations sometimes induce disorder. For example, the substrate binding domain of *E. coli* adenylate kinase was destabilized by the introduction of two point mutations [79]. Although locally unfolded, the resulting variant exhibited enhanced affinity for ATP and retained ca. 5% of the wild-type catalytic activity. These results suggest that the conformational switch between an open and closed state that normally

Table 2 Representative examples of inducibly disordered enzymes

Enzyme	Organism	Inducer	Structural disorder	Catalytic activity	Reference
Acylphosphatase	*S. solfataricus*	Guanidinium chloride	Partially disordered	79 wt%	[76]
Human topoisomerase I	*Homo sapiens*	Fragmentation/ recombination	Molten globule	~11 wt%	[77, 78]
I116G/L168G adenylate kinase	*E. coli*	Mutation	Partially disordered	5 wt%	[79]
S54G/P55N RNase T1	*Aspergillus oryzae*	Mutation	Partially disordered	40 wt%	[80]

limits the rate of catalysis occurs by cooperative unfolding and refolding of large segments of the protein. This novel molecular mechanism for achieving large-scale conformational transitions could conceivably be useful for other proteins as well.

A ribonuclease T1 mutant provides another example of a molten globule with substantial enzymatic activity. During refolding of an enzyme variant containing only a single *cis*-proline, a transient intermediate accumulates prior to proline isomerization that exhibits extensive secondary structure but only partial packing of the hydrophobic core. Despite its low tertiary structure content, this species possesses 40% of the RNase activity of the native protein toward the dinucleotide GpC [80]. Similarly, an intermediate containing a non-native peptidyl–prolyl bond is formed during refolding of RNase A that exhibits RNase activity similar to that of the native folded enzyme. However, subsequent studies showed that its active site region is largely in a native-like conformation [85]. In nature, kinetically trapped folding intermediates, which arise because of slow folding events like disulfide bond formation and proline isomerization, may also exhibit catalytic activity. Though not fully active, these species may contribute, at least to some extent, to biological function. These intrinsically disordered states might also shape the evolutionary potential of polypeptides by facilitating the generation of alternative folds and functions.

6 Designed Disordered Enzymes

While few instances of enzymatically active IDPs may be known in nature, a number of engineered proteins have been found to possess considerable catalytic power despite being intrinsically disordered (Table 3). In these cases, the sequence hallmarks of natural IDPs are not always evident.

The enzyme dihydrofolate reductase (DHFR) has been an important model system for studying protein folding, enzyme catalysis, and the relevance of protein dynamics for function [95]. It is a monomeric two-domain protein that catalyzes the reduction of 7,8-dihydrofolate (DHF) to 5,6,7,8-tetrahydrofolate (THF) by a

Table 3 Representative examples of designed enzymes that exhibit significant disorder

Enzyme	Organism	Modification	Structural disorder	Catalytic activity	Reference
DHFR (reductase)	*E. coli*	Circular permutation	Molten globule	1–5 wt%	[86]
C85A/C152S DHFR	*E. coli*	Fragmentation/ recombination	Molten globule	15 wt%	[87]
SNase (nuclease)	*S. aureus*	N-terminal truncation	Molten globule	65 wt%	[88]
F34W/W104F SNase	*S. aureus*	Mutation	Molten globule	87 wt%	[89]
N138D $SNase_{141}$	*S. aureus*	C-terminal truncation	Partially disordered	72 wt%	[90]
SNase	*S. aureus*	C-terminal truncation	Partially disordered	90–110 wt%	[91]
Δ131Δ SNase	*S. aureus*	C- and N-terminal truncation	Largely disordered	wt	[92]
TIM (isomerase)	–	Consensus design	Molten globule	0.002 wt%	[93]
CM (mutase)	*M. jannaschii*	Topological redesign	Molten globule	30 wt%	[94]

stereospecific hydride transfer from the cofactor NADPH to the C6 atom of the pterin ring of DHF (Fig. 3a). Five different ligand-complexed states in the catalytic cycle have been detected [95, 96]. These vary in the orientation and inherent flexibility of the Met20 loop, which governs access to the active site during catalysis [97]. In studies of chain connectivity effects on protein folding, the *E. coli* enzyme was extensively circularly permutated [87, 98–100]. The natural N- and C-termini were connected by a short peptide linker and new termini were sequentially introduced between every pair of protein residues (Fig. 3b). Certain permutations resulted in complete loss of structure and activity, while others had little or no effect. Two of the circularly permutated DHFR variants studied in detail possessed molten globular attributes but retained low catalytic activity, ranging from a 20- to 100-fold decrease in efficiency compared to wild-type DHFR [86]. Although they lack a rigid tertiary structure, addition of ligands like the potent inhibitor methotrexate leads to a gain of native-like structural properties, including cooperative thermal unfolding [101], suggesting that the small molecules may initiate folding. In an even more extreme modification, cleavage of the DHFR backbone after residue 86 produced two poorly structured, catalytically inactive fragments [87]. Nonetheless, when mixed, these formed a high affinity complex having substantially less secondary structure content than the wild-type protein but 15% of its DHFR activity. When exposed to inhibitors, the structural content of the complex increased significantly, indicating the potential of small molecules for stabilizing the active site architecture.

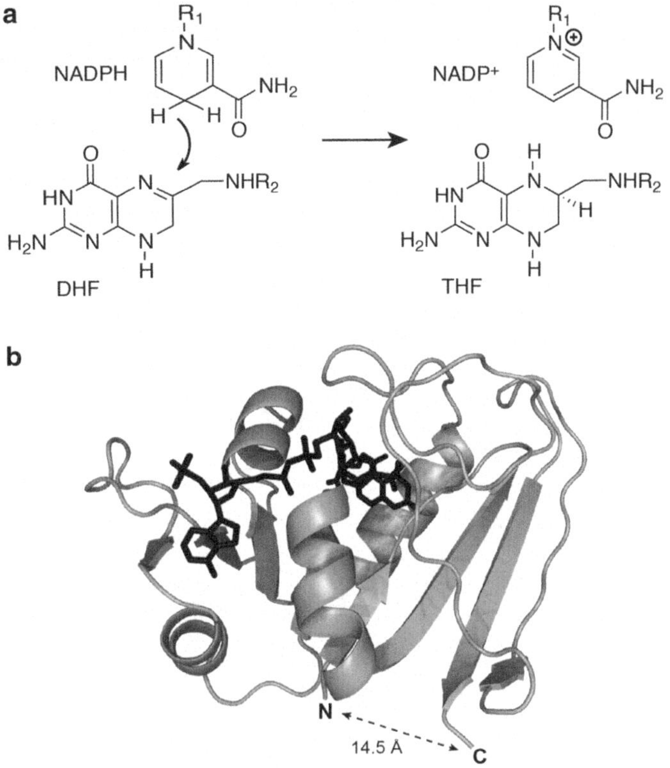

Fig. 3 Dihydrofolate reductase (DHFR). (**a**) The enzyme catalyzes stereospecific hydride transfer from NADPH to dihydrofolate (DHF) to give tetrahydrofolate (THF) and $NADP^+$. (**b**) Tertiary structure of DHFR (PDB code: 1RX2). The substrate folate and the cofactor NAD^+ are shown in *black*. The N- and C-termini of the protein are close in space, facilitating circular permutation

Diverse modifications of staphylococcal nuclease (SNase), a Ca^{2+}-dependent enzyme that catalyzes DNA and RNA hydrolysis, similarly produce intrinsically disordered but active enzyme variants (Fig. 4). For example, the double point mutant, F34W/W104F SNase, has molten globular-like characteristics and essentially wild-type catalytic activity [89]; in this case, ligand binding apparently causes the mutant to fold into its functionally active conformation. $SNase_{141}$, a truncated variant lacking the last eight C-terminal amino acids, has a compact, well-folded structure, but introduction of the N138D point mutation disrupts its conformational integrity and stability by eliminating a key hydrogen bond [90]. Nevertheless, this variant retains 72% of the activity of $SNase_{141}$. Another variant, whose last 13 amino acid residues were deleted, partly unfolds and lacks stable secondary structure [91, 102]. This variant is fully active at low salt concentrations. Calcium ions and the inhibitor thymidine 3′,5′-bisphosphate induce a conformational transformation to a more folded state [91].

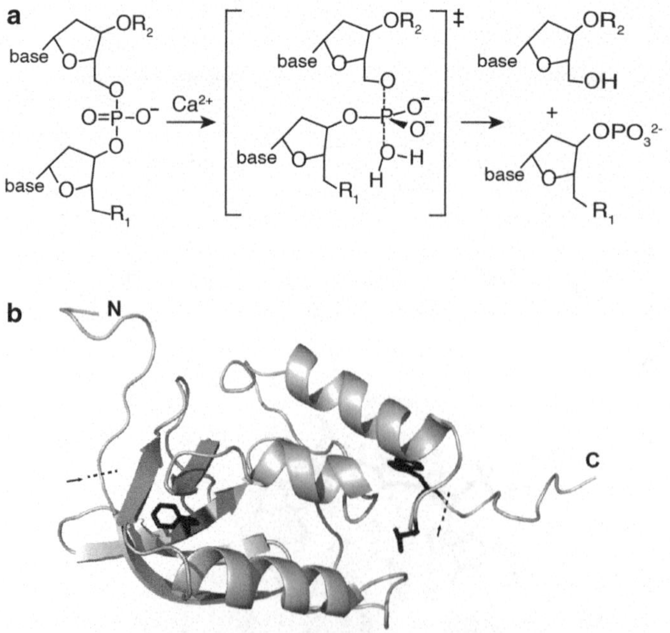

Fig. 4 Staphylococcal nuclease. (**a**) The enzyme is a calcium-dependent endonuclease that cleaves both single and double-stranded DNA and RNA. (**b**) Tertiary structure of wild-type staphylococcal nuclease (PDB code: 2SNS). Both N- and C-terminal truncations *(dashed lines)* produce intrinsically disordered protein. Residues Phe34 and Trp140, whose substitution to F34W/W140F results in disorder, are depicted in *black*. Residue Asn138, whose substitution to N138D combined with an eight amino acid C-terminal truncation results in disorder, is shown in *black*

Truncating the N-terminus of SNase also results in significant structural perturbations. Deleting 11 or 12 amino acids, for instance, produces a molten globular state [88, 103]. Whereas the −11 variant refolds to a distinct structure in the presence of substrate, −12 no longer adopts a native-like conformation. These variants possess 65% and 0.3% wild-type SNase activity, respectively. When residues 4–12 and 141–149 are excised, another variant is obtained – Δ131Δ SNase – whose native state is composed of different conformational species that rapidly convert between helical and extended conformations. Some secondary structure elements, such as the C-terminal α-helix or the fourth and fifth β-strands, are disordered in the native state. Nevertheless, Δ131Δ SNase assumes wild-type-like structures in the presence of the substrate and inhibitor, enabling efficient catalysis at high substrate concentrations [92, 104].

Studies on the capsid protease of Semliki Forest virus (SFVP) further illustrate the catalytic potential of IDPs [18]. Variants of this enzyme lacking one to seven C-terminal residues show biophysical properties characteristic of natively unfolded proteins, including the absence of a defined three-dimensional structure. The truncated proteins nevertheless efficiently catalyze the hydrolysis of activated aromatic amino acid esters with a 10^4-fold rate acceleration over background.

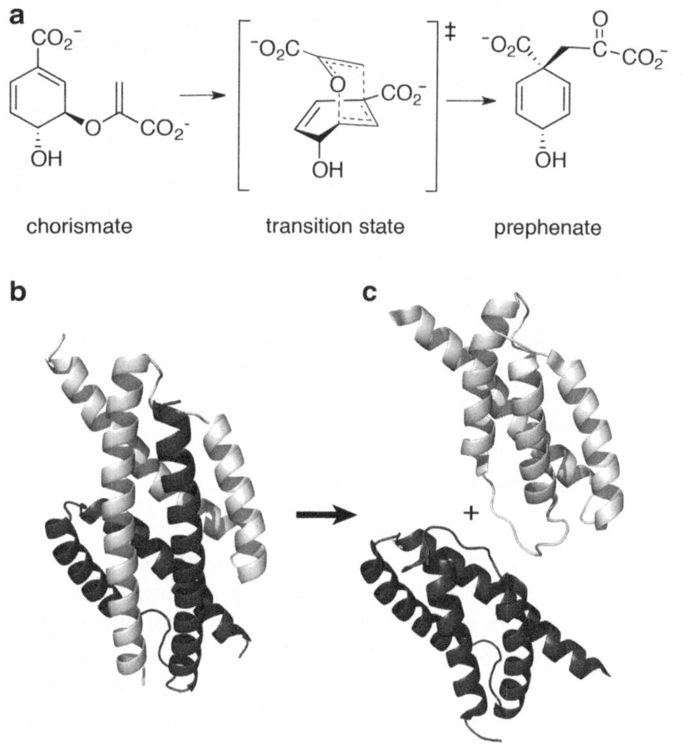

Fig. 5 Chorismate mutase. (**a**) The enzyme catalyzes the Claisen rearrangement of chorismate to prephenate via a pericyclic transition state. (**b**) Tertiary structure of a dimeric AroQ chorismate mutase (PDB code: 1ECM). (**c**) Insertion of a hinge-loop into the long N-terminal helix of structure of the dimeric AroQ mutase MjCM from *M. jannaschii* affords the highly active monomeric mutase mMjCM (PDB code: 2GTV), which exhibits the properties of a molten globule

Moreover, the reported k_{cat} and k_{cat}/K_m values of 15 s^{-1} and 5 × 10^5 M^{-1} s^{-1}, respectively, are comparable to the efficiencies of typical structured proteases.

In other work, a molten globular variant of triosephosphate isomerase (TIM) was inadvertently generated by consensus design [93]. The protein, cTIM, shares only 70% identity with its closest naturally occurring homolog. Biophysical characterization showed that it is monomeric, in contrast to native TIM dimers, and also less well folded. Although its catalytic efficiency (133 M^{-1} s^{-1}) is four orders of magnitude lower than that of the wild-type enzyme, this level of activity is sufficient to complement the TIM deficiency of an engineered *E. coli* strain. Interestingly, a relatively small number of mutations at unconserved positions sufficed to convert this consensus design into a protein with more native-like properties, including a well-folded structure and high catalytic efficiency.

One well-studied disordered enzyme is a designed chorismate mutase (CM) that catalyzes prephenate formation from chorismate, a key step in the biosynthesis of aromatic amino acids (Fig. 5). The homodimeric mutase from *Methanocaldococcus jannaschii* [105] was converted into a monomer by inserting a randomized hinge-loop sequence into the middle of the long, dimer-spanning N-terminal helix and selecting for functional variants in a CM-deficient *E. coli* strain [94]. The resulting protein, mMjCM, has 30% native activity but, unlike its thermostable parent, it possesses a fluctuating tertiary structure as judged by rapid H/D exchange, poor NMR signal dispersion, non-cooperative thermal denaturation, and ANS binding [106]. Upon binding of a transition state analog, this loosely packed and highly dynamic ensemble undergoes substantial structural ordering. This process is characterized by substantial entropy–enthalpy compensation as well as elevated rates of ligand association and dissociation compared to MjCM [107]. The solution structure of the complex, which was determined by NMR spectroscopy [24], confirmed that the protein adopts a bundle fold, as designed, although it still exhibits unprecedented millisecond flexibility across its entire length.

7 Protein Dynamics and Catalysis

The observation that structural disorder is compatible with catalysis is intriguing in the context of ongoing discussions regarding the role dynamics plays in the function of more conventional enzymes [108–114]. A variety of techniques, including fast kinetic methods, NMR relaxation data, alone or in combination with ambient-temperature X-ray crystallography, and molecular simulation techniques, have provided detailed information on the conformational dynamics along the reaction trajectories of selected enzymes [61, 96, 115, 116]. Conformational changes in otherwise well-structured proteins are often important for substrate binding and product release. By sequestering substrates from aqueous solution, structural rearrangements can create a protective environment for a reaction and position functional groups for effective transition state stabilization [117]. Sometimes conformational changes occur on a timescale similar to that of catalytic turnover and may even limit the overall rate of transformation [115, 118].

Attempts to link enzyme dynamics directly to catalytic efficiency – and particularly the ability of protein motions to influence the nature of the chemical step itself – have sparked vigorous debate [108–113, 116, 119–122]. Nevertheless, a consensus seems to be emerging that conformational sampling enhances the probability of generating active site environments that are conducive to reaction [109]. Protein fluctuations help to generate landscapes preorganized for molecular recognition of substrate(s) and stabilization of rate limiting transition states [123]. In so doing, they may help to reduce the reorganization energy required to move from the reactant to the transition state relative to the solution reaction. Nuclear quantum tunneling, which depends strongly on the hydrogen donor–acceptor distance, has been exploited to probe protein motions during catalysis of hydride transfer reactions [124]. The temperature dependence of the kinetic isotope effects, for example, has been cited as evidence for this type of conformational sampling [108]. By orienting the substrates for efficient reaction, protein motions influence the hydrogen donor–acceptor distance and create a suitable electrostatic environment

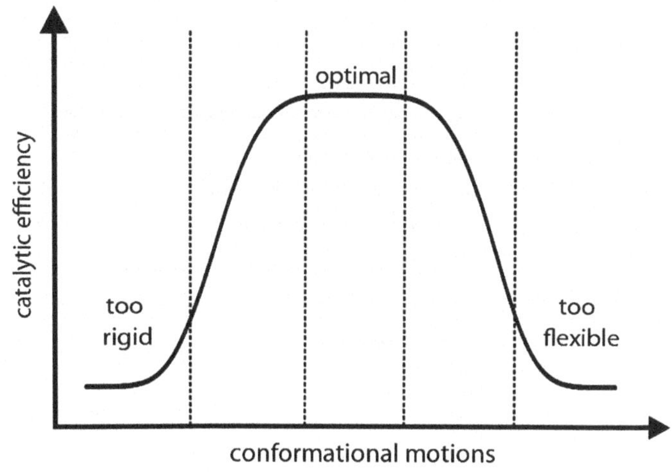

Fig. 6 Schematic representation of the relationship between protein flexibility and catalytic efficiency. Too high or too low inherent protein flexibility is expected to be disadvantageous for enzyme catalysis

for the reaction [113]. Such fluctuations are likely to be particularly important for multistep reactions that require continual active site reconfiguration to maximize chemical complementarity to the transition states of consecutive chemical steps. By funneling the protein through a series of catalytically competent conformations, a dynamic energy landscape can channel a reaction along a preferred kinetic path.

While some protein flexibility is clearly desirable to accommodate structural changes in the reactants as chemical bonds are made or broken, excessive structural disorder would be expected to diminish catalytic efficiency (Fig. 6). In fact, intrinsically disordered proteins – when enzymatically active – often exhibit modest catalytic proficiency compared to their well-folded counterparts [79, 86–88, 93]. Nevertheless, the turnover numbers for disordered enzymes, which vary from 10^{-2} to 10^{2} s^{-1} (Fig. 7), are comparable to the k_{cat} values reported for many natural enzymes [125]. Because the corresponding K_m parameters are in the micromolar to millimolar range, k_{cat}/K_m values can be as low as 100 M^{-1} s^{-1} or as high as 10^7 M^{-1} s^{-1} (Fig. 7). While these data are suggestive, relatively few disordered enzymes have been subject to detailed kinetic characterization, so only tentative conclusions regarding the influence of bound substrates on conformational landscapes can be made.

Structural heterogeneity should diminish the efficiency of conformational sampling. If the time required to generate productive conformations exceeds biologically relevant turnover times, which are typically in the millisecond to second range [126], the search for configurations capable of substrate binding could severely limit overall enzyme efficiency [127]. This may explain why many disordered catalysts found in nature exploit metal ions as organizing elements to nucleate and stabilize catalytically competent conformations.

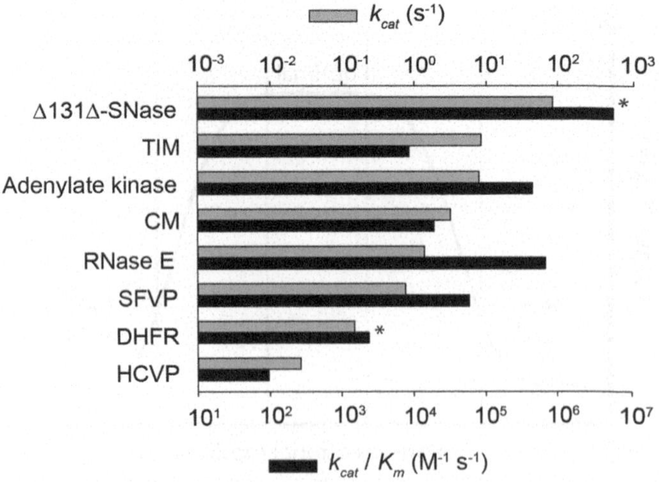

Fig. 7 Comparison of k_{cat} and $k_{\mathrm{cat}}/K_{\mathrm{m}}$ parameters for disordered enzymes that catalyze mechanistically diverse reactions. The *asterisk* denotes unpublished data from Stouffer and Schulenburg

The reduced activity observed for many disordered enzymes compared to their ordered counterparts may also reflect the difficulty of maintaining a productive catalytic environment for the time required for the chemical step to proceed [127]. The large number of charged amino acid residues could disfavor formation of a compact, catalytically competent fold even in the presence of bound ligand. The entropic cost associated with placing active site residues in a catalytically relevant conformation is likely to be high in disordered proteins, raising the free energy barrier for reaction and decreasing the overall rate. Nevertheless, tightening of non-covalent interactions throughout the complex upon ligand binding can give rise to enthalpic gains that (partially) offset such entropic losses [107]. In contrast, substantially less reorganization energy is required for the substrate to access transition state conformations in a well-folded and preorganized enzymatic active site.

As we have seen, some intrinsically disordered enzymes retain near native activity. These tend to be molten globules having largely intact secondary but fluctuating tertiary structures. In such cases, protein fluctuations are likely to represent modest departures from the mean structure observed for their well-folded homologs. As a consequence, locking the catalytically active conformation in place as the reaction proceeds has minimal energetic penalty. One case in point is the molten globular mMjCM chorismate mutase. Computational studies suggest that this enzyme undergoes greater ordering upon ligand binding than MjCM, its conventionally folded counterpart, but it also accesses a broader ensemble of catalytically competent conformations without significant preorganization penalty [128]. Consequently, the catalytic efficiency of the two systems differs by only a factor of three. Although conformational disorder decreases ligand affinity somewhat, mMjCM binds substrate and releases product significantly faster than MjCM [107]. In this case, partial structural disorder actually accelerates molecular recognition. The greater active site accessibility of the molten globule state probably facilitates formation of weak long-range protein–ligand interactions by a fly-casting mechanism, which strengthen as the enzyme accommodates to its target.

The proposal that chemical steps of an enzymatic reaction can be accelerated by channeling protein motions into vibrations along the reaction coordinate has generated considerable controversy. Experiments with "heavy" enzymes in which all nonexchangeable carbon, nitrogen, and hydrogen atoms were substituted with ^{13}C, ^{15}N, and ^{2}H have shown that a dynamic link might exist between mass-dependent bond vibrations of the enzyme and events in the reaction coordinate [129, 130]. However, the relevance of fast femtosecond to picosecond dynamic motions to catalysis is still debated. For example, the similar isotope effects on hydride transfer promoted by DHFR enzymes possessing widely different flexibilities argues against a direct coupling of protein motions to the chemical step [131]. Similarly, the fact that ordered and disordered proteins can both achieve high levels of enzymatic activity suggests that protein dynamics is not the source of the enormous rate acceleration over the reference solution reaction. The chorismate mutase mMjCM monomer exhibits motions on the same millisecond timescale as catalytic turnover [24], but it is difficult to envision how these stochastic fluctuations might be coupled dynamically to the chemical step. Instead, catalytically relevant active site configurations appear to be populated sufficiently frequently that little difference in rate is observed relative to the more ordered protein [128]. A recent computational study of cyclophilin A suggests that, while substrate dynamics associated with the chemical step of a reaction may be coupled to the dynamics of the surrounding medium, such effects do not lower the chemical barrier of an enzymatic reaction [132]. Instead, they alter the pre-exponential factor, effectively impeding the rate acceleration relative to the solution reaction.

Flexibility and cooperative conformational changes are general hallmarks of proteins generally and of enzymes more specifically. Most likely shaped by natural evolution, these dynamic properties vary widely across different scaffolds and even within the same protein fold family. For each specific case, there may be an ideal balance of conformational motion for optimal catalysis. However, establishing whether dynamical effects significantly influence the chemical step of an enzymatic reaction, and hence the reaction rates achieved, will require further research.

8 Structural Disorder and Evolution

Conformational dynamism has been proposed to be a cornerstone of protein evolvability [133–135]. Sampling scores of conformational options facilitates exploration of many potentially productive states. It may also favor the emergence of promiscuous binding and catalytic activities [49, 136, 137], which can be amplified in the course of divergent evolution [138]. Optimizing such activities through multiple rounds of mutagenesis and selection provides a route to new function.

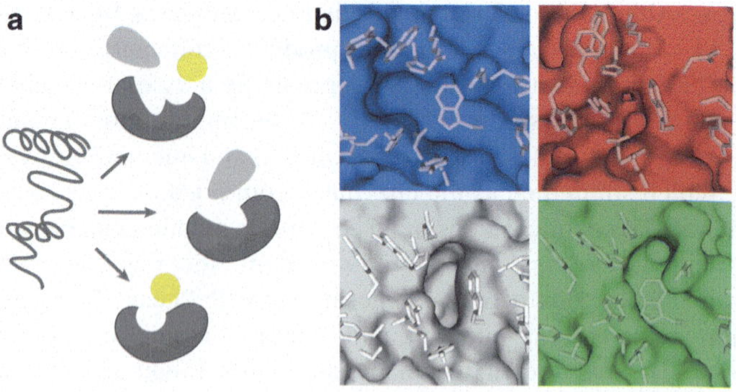

Fig. 8 Promiscuous functions of intrinsically disordered proteins. (**a**) Intrinsic disorder allows a protein to adopt different conformations, facilitating interactions with different binding partners. (**b**) Antibody SPE7 adopts multiple conformations in the absence of ligand (*blue*, PDB: 1OAQ and *red*, PDB: 1OCW), complexed with a low molecular weight hapten (*gray*, PDB: 1OAX), or bound to a protein (*green*, PDB: 1OAZ) [140]. The ligand has been omitted in the latter two panels to facilitate visualization of the respective binding pockets

Mammalian immune defense against diverse pathogens illustrates this point [139]. In response to antigenic challenge, antibody receptors are produced that bind foreign molecules with high affinity and selectivity. An enormously diverse pool of immunoglobulin sequences are initially assembled at the genetic level by combining multiple gene segments. Structural studies on germ-line antibodies show that the proteins in the primary immunological repertoire often exhibit high conformational diversity. The effective scope of the starting immunological repertoire is further increased by the ability of individual binding pockets to adapt to many different antigens, providing countless viable starting points for the evolution of more stable and selective complexes via somatic mutation and affinity selection. Over the course of affinity maturation, mutations that stabilize specific conformations from the ensemble of possibilities are selected, converting a low-affinity, imperfect binding site to a high-affinity complex with tailored binding interactions [135]. In some instances, even mature antibodies have been shown to adopt multiple structures and bind structurally unrelated antigens (Fig. 8) [140].

While all proteins and their functions are subject to evolutionary forces, disordered proteins should be easier to alter and adapt for new tasks than their fully structured counterparts. Individual amino acids in flexible segments take on diverse conformations more freely, and are less constrained by structure, enabling successful accumulation of neutral mutations and emergence of new function. Further facilitating evolution in IDPs is their large fraction of solvent exposed residues; surface residues are generally subject to more frequent substitution than amino acids in the buried protein core [141]. Indeed, sequence variability correlates with protein packing density [142], implying that loosely packed proteins evolve more rapidly than proteins with high tertiary structural content. By comparing genetic distance in protein families, proteins having sizable disordered regions were found to have evolved significantly faster than ordered proteins in 19 out of 26 instances [143]. In only two cases did disordered proteins evolve more slowly than their ordered counterparts.

In a direct experimental test of connections between active site flexibility and evolvability, mutational tolerance of the intrinsically disordered chorismate mutase mMjCM was compared to that of a homologous thermostable enzyme [144]. It was found that analogous point mutations in ordered and disordered scaffolds have widely divergent and context-dependent effects on catalytic activity. Because variations in activity were not caused by decreases in secondary structure or thermodynamic stability, the results substantiate the idea that enzymes lose their activities more easily than their folds upon mutation. Less stable structures such as molten globules are evidently viable starting points for evolution. In fact, when the molten globular mMjCM was subjected to three rounds of random mutagenesis and high-stringency selection, a robust and more native-like variant was obtained that exhibited catalytic efficiency comparable to that of a natural dimeric chorismate mutase [20]. Biophysical characterization demonstrated that the evolved enzymes were substantially better folded and more stable than their molten globular precursor.

While such findings support the premise that disordered structures are promising starting points for evolution, a pre-existing activity was optimized in the mMjCM experiments. Changing the catalytic mechanism of an enzyme represents a far greater challenge. The catalytic promiscuity that many enzymes exhibit provides a possible handle for redesigning substrate specificity and catalytic mechanism by directed evolution. Several reports describe the optimization of such activities by random mutagenesis and screening [145–151], but it remains to be seen whether high structural plasticity further enhances evolvability. Although it should be easier for a conformationally diverse ensemble of molecules to access new catalytic activity than a conventionally folded protein, small structural changes could further destabilize the molecule, resulting in complete unfolding or faster degradation. Preliminary attempts to convert the molten globular chorismate mutase into an isochorismate pyruvate lyase have proven surprisingly difficult, for example (K. Höland, K.J. Woycechowsky, C. Jäckel, and D. Hilvert, unpublished data). Despite the likelihood that some natural isochorismate pyruvate lyases evolved from AroQ mutases [152], and clear sequence, structural, and chemical homology between the two systems, it has not been possible to achieve the desired functional transformation through mutagenesis and screening. Improved understanding of the interplay between protein disorder, promiscuity, and evolvability will hopefully emerge from additional study.

Many, if not most, modern enzymes are believed to derive from primordial molten globules [153, 154]. Given this, the design of new enzymes in the laboratory may also profit from intrinsic disorder. Protein engineers have created many de novo folds that possess the properties of molten globules; these could serve as the starting point for novel biochemical activities.

9 Perspectives

Over the past two decades, numerous IDPs have been described in the biochemical literature. Due to high flexibility and enlarged interaction surfaces, these proteins possess significant advantages over fully folded proteins for specific biological tasks. While high specificity and moderate binding affinity enable interaction with many different binding partners, increased cellular turnover also enables rapid response at fundamental points of the cellular signaling network [155]. The ability to perform many biologically important activities outside the native state is evidently more prevalent than previously thought. That enzymes too can also be substantially disordered is, surely, still something of a surprise.

The fact that some disordered proteins achieve substantial levels of catalytic activity challenges the long-held view that efficient catalysis requires a high degree of structural preorganization. Though conformational heterogeneity might be expected to reduce catalytic efficiency, it is now clear that the energetic cost of forming a suitable environment for a chemical reaction need not be terribly high. The disorder-to-order transition observed for intrinsically disordered enzymes upon ligand binding is analogous to the conformational sampling and reorganization processes that also take place in most conventionally folded enzymes, differing perhaps only in degree.

The finding that catalysis is not necessarily coupled to a stable and persistent fold nevertheless opens new perspectives on enzyme action and promises to enhance our understanding of the role of dynamics on normal enzyme function. From a practical viewpoint, intrinsic disorder may provide some enzymes with a novel selectable advantage in much the same way as it extends the properties of protein receptors and interaction partners. Rapid ligand binding and release with only modest losses in affinity can ensure rapid flux through the catalyst and avoidance of product inhibition. The ability to sample multiple catalytically relevant configurations could be especially helpful for multi-step reactions that require continuous reorganization of the binding pocket to accommodate consecutive transition states. Tight regulation of the production, posttranslational modification, and degradation of these catalysts, as for other disordered proteins, should allow rapid adjustment of intracellular concentrations as needed. Promiscuous binding to structurally distinct substrate molecules provides a means of rapid evolutionary adaptation. In the future it may be possible to exploit these properties for the purposes of design.

Acknowledgments The authors are grateful to Dr. An Vandemeulebroucke and Dr. Vladimir Torbeev for fruitful discussions and carefully reading the manuscript and Richard Obexer for refining the figures. Investigations of disordered proteins in the Hilvert group have been generously supported by the Schweizerischer Nationalfonds and the ETH Zürich.

References

1. Uversky VN (2011) Intrinsically disordered proteins from A to Z. Int J Biochem Cell Biol 43:1090–1103
2. Fisher CK, Stultz CM (2011) Constructing ensembles for intrinsically disordered proteins. Curr Opin Struct Biol 21:426–431
3. Tompa P, Dosztanyi Z, Simon I (2006) Prevalent structural disorder in *E. coli* and *S. cerevisiae* proteomes. J Proteome Res 5:1996–2000
4. He B, Wang K, Liu Y, Xue B, Uversky VN, Dunker AK (2009) Predicting intrinsic disorder in proteins: an overview. Cell Res 19:929–949
5. Tompa P (2012) Intrinsically disordered proteins: a 10-year recap. Trends Biochem Sci. 37: 509–516
6. Dunker AK, Silman I, Uversky VN, Sussman JL (2008) Function and structure of inherently disordered proteins. Curr Opin Struct Biol 18:756–764
7. Uversky VN, Oldfield CJ, Dunker AK (2005) Showing your ID: intrinsic disorder as an ID for recognition, regulation and cell signaling. J Mol Recognit 18:343–384
8. Henzler-Wildman K, Kern D (2007) Dynamic personalities of proteins. Nature 450:964–972
9. Karplus M, McCammon JA (1981) The internal dynamics of globular proteins. CRC Crit Rev Biochem 9:293–349
10. Henzler-Wildman KA, Lei M, Thai V, Kerns SJ, Karplus M, Kern D (2007) A hierarchy of timescales in protein dynamics is linked to enzyme catalysis. Nature 450:913–916
11. Dunker AK, Oldfield CJ, Meng J, Romero P, Yang JY, Chen JW, Vacic V, Obradovic Z, Uversky VN (2008) The unfoldomics decade: an update on intrinsically disordered proteins. BMC Genomics 9(Suppl 2):S1
12. Romero P, Obradovic Z, Li XH, Garner EC, Brown CJ, Dunker AK (2001) Sequence complexity of disordered protein. Proteins 42:38–48
13. Fitzkee NC, García-Moreno EB (2008) Electrostatic effects in unfolded staphylococcal nuclease. Protein Sci 17:216–227
14. Mao AH, Crick SL, Vitalis A, Chicoine CL, Pappu RV (2010) Net charge per residue modulates conformational ensembles of intrinsically disordered proteins. Proc Natl Acad Sci USA 107:8183–8188
15. Müller-Späth S, Soranno A, Hirschfeld V, Hofmann H, Rüegger S, Reymond L, Nettels D, Schuler B (2010) Charge interactions can dominate the dimensions of intrinsically disordered proteins. Proc Natl Acad Sci USA 107:14609–14614
16. Uversky VN, Gillespie JR, Fink AL (2000) Why are "natively unfolded" proteins unstructured under physiologic conditions? Proteins 41:415–427
17. Zambelli B, Musiani F, Benini S, Ciurli S (2011) Chemistry of Ni^{2+} in urease: sensing, trafficking, and catalysis. Acc Chem Res 44:520–530
18. Morillas M, Eberl H, Allain FH-T, Glockshuber R, Kuennemann E (2008) Novel enzymatic activity derived from the Semliki Forest virus capsid protein. J Mol Biol 376:721–735
19. Abian O, Vega S, Neira JL, Velazquez-Campoy A (2010) Conformational stability of hepatitis C virus NS3 protease. Biophys J 99:3811–3820
20. Butz M, Neuenschwander M, Kast P, Hilvert D (2011) An N-terminal protein degradation tag enables robust selection of highly active enzymes. Biochemistry 50:8594–8602
21. Cardamone M, Puri NK (1992) Spectrofluorimetric assessment of the surface hydrophobicity of proteins. Biochem J 282(Pt 2):589–593
22. Receveur-Bréchot V, Durand D (2011) How random are intrinsically disordered proteins? A small angle scattering perspective. Curr Protein Peptide Sci 13:55–75
23. Bernadó P, Modig K, Grela P, Svergun DI, Tchorzewski M, Pons M, Akke M (2010) Structure and dynamics of ribosomal protein L12: an ensemble model based on SAXS and NMR relaxation. Biophys J 98:2374–2382
24. Pervushin K, Vamvaca K, Vögeli B, Hilvert D (2007) Structure and dynamics of a molten globular enzyme. Nat Struct Mol Biol 14:1202–1206

25. Mittag T, Forman-Kay JD (2007) Atomic-level characterization of disordered protein ensembles. Curr Opin Struct Biol 17:3–14
26. Bernadó P, Svergun DI (2011) Structural analysis of intrinsically disordered proteins by small-angle X-ray scattering. Mol Biosyst 8:151–167
27. Zambelli B, Stola M, Musiani F, De Vriendt K, Samyn B, Devreese B, Van Beeumen J, Turano P, Dikiy A, Bryant DA et al (2005) UreG, a chaperone in the urease assembly process, is an intrinsically unstructured GTPase that specifically binds Zn^{2+}. J Biol Chem 280: 4684–4695
28. Zotter Á, Oláh J, Hlavanda E, Bodor A, Perczel A, Szigeti K, Fidy J, Ovádi J (2011) Zn^{2+}-induced rearrangement of the disordered TPPP/p25 affects its microtubule assembly and GTPase activity. Biochemistry 50:9568–9578
29. Fitzkee NC, Masse JE, Shen Y, Davies DR, Bax A (2010) Solution conformation and dynamics of the HIV-1 integrase core domain. J Biol Chem 285:18072–18084
30. Fitzkee NC, Torchia DA, Bax A (2011) Measuring rapid hydrogen exchange in the homodimeric 36 kDa HIV-1 integrase catalytic core domain. Protein Sci 20:500–512
31. Galea CA, Wang Y, Sivakolundu SG, Kriwacki RW (2008) Regulation of cell division by intrinsically unstructured proteins: intrinsic flexibility, modularity, and signaling conduits. Biochemistry 47:7598–7609
32. Wright PE, Dyson HJ (1999) Intrinsically unstructured proteins: re-assessing the protein structure-function paradigm. J Mol Biol 293:321–331
33. Tantos A, Han KH, Tompa P (2012) Intrinsic disorder in cell signaling and gene transcription. Mol Cell Endocrinol 348:457–465
34. Uversky VN, Oldfield CJ, Dunker AK (2008) Intrinsically disordered proteins in human diseases: introducing the D2 concept. Annu Rev Biophys 37:215–246
35. Chakrabortee S, Meersman F, Kaminski Schierle GS, Bertoncini CW, McGee B, Kaminski CF, Tunnacliffe A (2010) Catalytic and chaperone-like functions in an intrinsically disordered protein associated with desiccation tolerance. Proc Natl Acad Sci USA 107: 16084–16089
36. Reichmann D, Xu Y, Cremers CM, Ilbert M, Mittelman R, Fitzgerald MC, Jakob U (2012) Order out of disorder: working cycle of an intrinsically unfolded chaperone. Cell 148: 947–957
37. Tompa P, Csermely P (2004) The role of structural disorder in the function of RNA and protein chaperones. FASEB J 18:1169–1175
38. Wright PE, Dyson HJ (2009) Linking folding and binding. Curr Opin Struct Biol 19:31–38
39. Fuxreiter M, Simon I, Friedrich P, Tompa P (2004) Preformed structural elements feature in partner recognition by intrinsically unstructured proteins. J Mol Biol 338:1015–1026
40. Tsai CD, Ma B, Kumar S, Wolfson H, Nussinov R (2001) Protein folding: binding of conformationally fluctuating building blocks via population selection. Crit Rev Biochem Mol Biol 36:399–433
41. Kiefhaber T, Bachmann A, Jensen KS (2012) Dynamics and mechanisms of coupled protein folding and binding reactions. Curr Opin Struct Biol 22:21–29
42. Receveur-Brechot V, Bourhis JM, Uversky VN, Canard B, Longhi S (2006) Assessing protein disorder and induced folding. Proteins 62:24–45
43. Borriello A, Cucciolla V, Oliva A, Zappia V, Della Ragione F (2007) p27Kip1 metabolism: a fascinating labyrinth. Cell Cycle 6:1053–1061
44. Lowry DF, Hausrath AC, Daughdrill GW (2008) A robust approach for analyzing a heterogeneous structural ensemble. Proteins 73:918–928
45. Sugase K, Dyson HJ, Wright PE (2007) Mechanism of coupled folding and binding of an intrinsically disordered protein. Nature 447:1021–1025
46. Turjanski AG, Gutkind JS, Best RB, Hummer G (2008) Binding-induced folding of a natively unstructured transcription factor. PLoS Comput Biol 4:e1000060
47. Espinoza-Fonseca LM (2009) Reconciling binding mechanisms of intrinsically disordered proteins. Biochem Biophys Res Commun 382:479–482

48. Oldfield CJ, Meng J, Yang JY, Yang MQ, Uversky VN, Dunker AK (2008) Flexible nets: disorder and induced fit in the associations of p53 and 14-3-3 with their partners. BMC Genomics 9(Suppl 1):S1
49. Schreiber G, Keating AE (2011) Protein binding specificity versus promiscuity. Curr Opin Struct Biol 21:50–61
50. Taira N, Yoshida K (2012) Post-translational modifications of p53 tumor suppressor: determinants of its functional targets. Histol Histopathol 27:437–443
51. Vuzman D, Levy Y (2012) Intrinsically disordered regions as affinity tuners in protein-DNA interactions. Mol Biosyst 8:47–57
52. Shoemaker BA, Portman JJ, Wolynes PG (2000) Speeding molecular recognition by using the folding funnel: the fly-casting mechanism. Proc Natl Acad Sci USA 97:8868–8873
53. Huang Y, Liu Z (2009) Kinetic advantage of intrinsically disordered proteins in coupled folding–binding process: a critical assessment of the "fly-casting" mechanism. J Mol Biol 393:1143–1159
54. Fink AL (2005) Natively unfolded proteins. Curr Opin Struct Biol 15:35–41
55. Patil A, Kinoshita K, Nakamura H (2010) Domain distribution and intrinsic disorder in hubs in the human protein–protein interaction network. Protein Sci 19:1461–1468
56. Singh GP, Ganapathi M, Dash D (2007) Role of intrinsic disorder in transient interactions of hub proteins. Proteins 66:761–765
57. Dunker AK, Cortese MS, Romero P, Iakoucheva LM, Uversky VN (2005) Flexible nets. The roles of intrinsic disorder in protein interaction networks. FEBS J 272:5129–5148
58. Pauling L (1948) Nature of forces between large molecules of biological interest. Nature 161: 707–709
59. Koshland DE (1958) Application of a theory of enzyme specificity to protein synthesis. Proc Natl Acad Sci USA 44:98–104
60. Koshland DE (1994) The key-lock theory and the induced fit theory. Angew Chem Int Ed Engl 33:2375–2378
61. Hammes GG, Benkovic SJ, Hammes-Schiffer S (2011) Flexibility, diversity, and cooperativity: pillars of enzyme catalysis. Biochemistry 50:10422–10430
62. Redko Y, Tock MR, Adams CJ, Kaberdin VR, Grasby JA, McDowall KJ (2003) Determination of the catalytic parameters of the N-terminal half of *Escherichia coli* ribonuclease E and the identification of critical functional groups in RNA substrates. J Biol Chem 278: 44001–44008
63. Zhang RM, Durkin J, Windsor WT, McNemar C, Ramanathan L, Le HV (1997) Probing the substrate specificity of hepatitis C virus NS3 serine protease by using synthetic peptides. J Virol 71:6208–6213
64. Zambelli B, Musiani F, Savini M, Tucker P, Ciurli S (2007) Biochemical studies on *Mycobacterium tuberculosis* UreG and comparative modeling reveal structural and functional conservation among the bacterial UreG family. Biochemistry 46:3171–3182
65. Kaberdin VR, Miczak A, Jakobsen JS, Lin-Chao S, McDowall KJ, von Gabain A (1998) The endoribonucleolytic N-terminal half of *Escherichia coli* RNase E is evolutionarily conserved in *Synechocystis sp.* and other bacteria but not the C-terminal half, which is sufficient for degradosome assembly. Proc Natl Acad Sci USA 95:11637–11642
66. Callaghan AJ, Aurikko JP, Ilag LL, Günter Grossmann J, Chandran V, Kühnel K, Poljak L, Carpousis AJ, Robinson CV, Symmons MF et al (2004) Studies of the RNA degradosome-organizing domain of the *Escherichia coli* ribonuclease RNase E. J Mol Biol 340:965–979
67. Carpousis AJ (2007) The RNA degradosome of *Escherichia coli*: an mRNA-degrading machine assembled on RNase E. Annu Rev Microbiol 61:71–87
68. Tedbury PR, Harris M (2007) Characterisation of the role of zinc in the hepatitis C virus NS2/3 auto-cleavage and NS3 protease activities. J Mol Biol 366:1652–1660
69. Zambelli B, Cremades N, Neyroz P, Turano P, Uversky VN, Ciurli S (2012) Insights in the (un)structural organization of *Bacillus pasteurii* UreG, an intrinsically disordered GTPase enzyme. Mol Biosyst 8:220–228

70. Patil A, Kinoshita K, Nakamura H (2010) Hub promiscuity in protein–protein interaction networks. Int J Mol Sci 11:1930–1943
71. Tirián L, Hlavanda E, Oláh J, Horváth I, Orosz F, Szabó B, Kovács J, Szabad J, Ovádi J (2003) TPPP/p25 promotes tubulin assemblies and blocks mitotic spindle formation. Proc Natl Acad Sci USA 100:13976–13981
72. Ovádi J, Orosz F (2009) An unstructured protein with destructive potential: TPPP/p25 in neurodegeneration. Bioessays 31:676–686
73. Hlavanda E, Kovács J, Oláh J, Orosz F, Medzihradszky KF, Ovádi J (2002) Brain-specific p25 protein binds to tubulin and microtubules and induces aberrant microtubule assemblies at substoichiometric concentrations. Biochemistry 41:8657–8664
74. Kovács GG, László L, Kovács J, Jensen PH, Lindersson E, Botond G, Molnár T, Perczel A, Hudecz F, Mezo G et al (2004) Natively unfolded tubulin polymerization promoting protein TPPP/p25 is a common marker of alpha-synucleinopathies. Neurobiol Dis 17:155–162
75. Fink AL, Calciano LJ, Goto Y, Kurotsu T, Palleros DR (1994) Classification of acid denaturation of proteins: intermediates and unfolded states. Biochemistry 33:12504–12511
76. Bemporad F, Gsponer J, Hopearuoho HI, Plakoutsi G, Stati G, Stefani M, Taddei N, Vendruscolo M, Chiti F (2008) Biological function in a non-native partially folded state of a protein. EMBO J 27:1525–1535
77. Punchihewa C, Dai J, Carver M, Yang D (2007) Human topoisomerase I C-terminal domain fragment containing the active site tyrosine is a molten globule: implication for the formation of competent productive complex. J Struct Biol 159:111–121
78. Stewart L, Ireton GC, Champoux JJ (1997) Reconstitution of human topoisomerase I by fragment complementation. J Mol Biol 269:355–372
79. Olsson U, Wolf-Watz M (2010) Overlap between folding and functional energy landscapes for adenylate kinase conformational change. Nat Commun 1:111
80. Kiefhaber T, Schmid FX, Willaert K, Engelborghs Y, Chaffotte A (1992) Structure of a rapidly formed intermediate in ribonuclease T1 folding. Protein Sci 1:1162–1172
81. Ptitsyn OB, Pain RH, Semisotnov GV, Zerovnik E, Razgulyaev OI (1990) Evidence for a molten globule state as a general intermediate in protein folding. FEBS Lett 262:20–24
82. Goldberg ME, Semisotnov GV, Friguet B, Kuwajima K, Ptitsyn OB, Sugai S (1990) An early immunoreactive folding intermediate of the tryptophan synthease beta 2 subunit is a 'molten globule'. FEBS Lett 263:51–56
83. Creighton TE (1997) How important is the molten globule for correct protein folding? Trends Biochem Sci 22:6–10
84. Hu J, Li D, Su X-D, Jin C, Xia B (2010) Solution structure and conformational heterogeneity of acylphosphatase from *Bacillus subtilis*. FEBS Lett 584:2852–2856
85. Schmid FX, Blaschek H (1981) A native-like intermediate on the ribonuclease A folding pathway. 2. Comparison of its properties to native ribonuclease A. Eur J Biochem 114: 111–117
86. Protasova NY, Kireeva ML, Murzina NV, Murzin AG, Uversky VN, Gryaznova OI, Gudkov AT (1994) Circularly permuted dihydrofolate-reductase of *Escherichia coli* has functional activity and a destabilized tertiary structure. Protein Eng 7:1373–1377
87. Smith VF, Matthews CR (2001) Testing the role of chain connectivity on the stability and structure of dihydrofolate reductase from *E. coli*: fragment complementation and circular permutation reveal stable, alternatively folded forms. Protein Sci 10:116–128
88. Zhang H, Huang S, Feng Y, Guo P, Jing G (2005) Effect of N-terminal deletions on the foldability, stability, and activity of staphylococcal nuclease. Arch Biochem Biophys 441: 123–131
89. Li Y, Jing G (2000) Double point mutant F34W/W140F of staphylococcal nuclease is in a molten globule state but highly competent to fold into a functional conformation. J Biochem 128:739–744

90. Huang S, Yin J, Feng Y, Jing G (2003) Effect of a specific hydrogen bond (N138ND2-Q106O) on conformational integrity, stability, and activity of staphylococcal nuclease. Arch Biochem Biophys 420:87–94
91. Shortle D, Meeker AK (1989) Residual structure in large fragments of staphylococcal nuclease: effects of amino acid substitutions. Biochemistry 28:936–944
92. Alexandrescu AT, Dames SA, Wiltscheck R (1996) A fragment of staphylococcal nuclease with an OB-fold structure shows hydrogen-exchange protection factors in the range reported for "molten globules". Protein Sci 5:1942–1946
93. Sullivan BJ, Durani V, Magliery TJ (2011) Triosephosphate isomerase by consensus design: dramatic differences in physical properties and activity of related variants. J Mol Biol 413: 195–208
94. MacBeath G, Kast P, Hilvert D (1998) Redesigning enzyme topology by directed evolution. Science 279:1958–1961
95. Schnell JR, Dyson HJ, Wright PE (2004) Structure, dynamics, and catalytic function of dihydrofolate reductase. Annu Rev Biophys Biomol Struct 33:119–140
96. Boehr DD, McElheny D, Dyson HJ, Wright PE (2006) The dynamic energy landscape of dihydrofolate reductase catalysis. Science 313:1638–1642
97. Sawaya MR, Kraut J (1997) Loop and subdomain movements in the mechanism of *Escherichia coli* dihydrofolate reductase: crystallographic evidence. Biochemistry 36: 586–603
98. Iwakura M (1998) In search of circular permuted variants of *Escherichia coli* dihydrofolate reductase. Biosci Biotechnol Biochem 62:778–781
99. Hu Z, Bowen D, Southerland WM, del Sol A, Pan Y, Nussinov R, Ma B (2007) Ligand binding and circular permutation modify residue interaction network in DHFR. PLoS Comput Biol 3:e117
100. Svensson A-KE, Zitzewitz JA, Matthews CR, Smith VF (2006) The relationship between chain connectivity and domain stability in the equilibrium and kinetic folding mechanisms of dihydrofolate reductase from *E. coli*. Protein Eng Des Sel 19:175–185
101. Uversky VN, Kutyshenko VP, Protasova NY, Rogov VV, Vassilenko KS, Gudkov AT (1996) Circularly permuted dihydrofolate reductase possesses all the properties of the molten globule state, but can resume functional tertiary structure by interaction with its ligands. Protein Sci 5:1844–1851
102. Flanagan JM, Kataoka M, Shortle D, Engelman DM (1992) Truncated staphylococcal nuclease is compact but disordered. Proc Natl Acad Sci USA 89:748–752
103. Ermácora MR, Ledman DW, Fox RO (1996) Mapping the structure of a non-native state of staphylococcal nuclease. Nat Struct Biol 3:59–66
104. Alexandrescu AT, Jahnke W, Wiltscheck R, Blommers MJ (1996) Accretion of structure in staphylococcal nuclease: an 15N NMR relaxation study. J Mol Biol 260:570–587
105. MacBeath G, Kast P, Hilvert D (1998) A small, thermostable, and monofunctional chorismate mutase from the archaeon *Methanococcus jannaschii*. Biochemistry 37: 10062–10073
106. Vamvaca K, Vögeli B, Kast P, Pervushin K, Hilvert D (2004) An enzymatic molten globule: efficient coupling of folding and catalysis. Proc Natl Acad Sci USA 101:12860–12864
107. Vamvaca K, Jelesarov I, Hilvert D (2008) Kinetics and thermodynamics of ligand binding to a molten globular enzyme and its native counterpart. J Mol Biol 382:971–977
108. Nagel ZD, Klinman JP (2009) A 21st century revisionist's view at a turning point in enzymology. Nat Chem Biol 5:543–550
109. Nashine VC, Hammes-Schiffer S, Benkovic SJ (2010) Coupled motions in enzyme catalysis. Curr Opin Chem Biol 14:644–651
110. Bhabha G, Lee J, Ekiert DC, Gam J, Wilson IA, Dyson HJ, Benkovic SJ, Wright PE (2011) A dynamic knockout reveals that conformational fluctuations influence the chemical step of enzyme catalysis. Science 332:234–238

111. Adamczyk AJ, Cao J, Kamerlin SCL, Warshel A (2011) Catalysis by dihydrofolate reductase and other enzymes arises from electrostatic preorganization, not conformational motions. Proc Natl Acad Sci USA 108:14115–14120
112. Henzler-Wildman KA, Thai V, Lei M, Ott M, Wolf-Watz M, Fenn T, Pozharski E, Wilson MA, Petsko GA, Karplus M et al (2007) Intrinsic motions along an enzymatic reaction trajectory. Nature 450:838–844
113. Warshel A, Sharma PK, Kato M, Xiang Y, Liu H, Olsson MHM (2006) Electrostatic basis for enzyme catalysis. Chem Rev 106:3210–3235
114. Loveridge EJ, Behiry EM, Guo J, Allemann RK (2012) Evidence that a 'dynamic knockout' in *Escherichia coli* dihydrofolate reductase does not affect the chemical step of catalysis. Nat Chem 4:292–297
115. Palmer AG (2004) NMR characterization of the dynamics of biomacromolecules. Chem Rev 104:3623–3640
116. Eisenmesser EZ, Millet O, Labeikovsky W, Korzhnev DM, Wolf-Watz M, Bosco DA, Skalicky JJ, Kay LE, Kern D (2005) Intrinsic dynamics of an enzyme underlies catalysis. Nature 438:117–121
117. Schramm VL (1998) Enzymatic transition states and transition state analog design. Annu Rev Biochem 67:693–720
118. Torbeev VY, Raghuraman H, Hamelberg D, Tonelli M, Westler WM, Perozo E, Kent SB (2011) Protein conformational dynamics in the mechanism of HIV-1 protease catalysis. Proc Natl Acad Sci USA 108:20982–20987
119. Garcia-Viloca M, Gao J, Karplus M, Truhlar DG (2004) How enzymes work: analysis by modern rate theory and computer simulations. Science 303:186–195
120. Pineda JRET, Antoniou D, Schwartz SD (2010) Slow conformational motions that favor sub-picosecond motions important for catalysis. J Phys Chem B 114:15985–15990
121. Pisliakov AV, Cao J, Kamerlin SCL, Warshel A (2009) Enzyme millisecond conformational dynamics do not catalyze the chemical step. Proc Natl Acad Sci USA 106:17359–17364
122. Schramm VL (2011) Enzymatic transition states, transition-state analogs, dynamics, thermodynamics, and lifetimes. Annu Rev Biochem 80:703–732
123. Rajagopalan PT, Benkovic SJ (2002) Preorganization and protein dynamics in enzyme catalysis. Chem Rec 2:24–36
124. Nagel ZD, Klinman JP (2006) Tunneling and dynamics in enzymatic hydride transfer. Chem Rev 106:3095–3118
125. Bar-Even A, Noor E, Savir Y, Liebermeister W, Davidi D, Tawfik DS, Milo R (2011) The moderately efficient enzyme: evolutionary and physicochemical trends shaping enzyme parameters. Biochemistry 50:4402–4410
126. Wolfenden R, Snider MJ (2001) The depth of chemical time and the power of enzymes as catalysts. Acc Chem Res 34:938–945
127. Vendruscolo M (2010) Enzymatic activity in disordered states of proteins. Curr Opin Chem Biol 14:671–675
128. Roca M, Messer B, Hilvert D, Warshel A (2008) On the relationship between folding and chemical landscapes in enzyme catalysis. Proc Natl Acad Sci USA 105:13877–13882
129. Silva RG, Murkin AS, Schramm VL (2011) Femtosecond dynamics coupled to chemical barrier crossing in a Born–Oppenheimer enzyme. Proc Natl Acad Sci USA 108:18661–18665
130. Kipp DR, Silva RG, Schramm VL (2011) Mass-dependent bond vibrational dynamics influence catalysis by HIV-1 protease. J Am Chem Soc 133:19358–19361
131. Loveridge EJ, Tey L-H, Behiry EM, Dawson WM, Evans RM, Whittaker SB-M, Günther UL, Williams C, Crump MP, Allemann RK (2011) The role of large-scale motions in catalysis by dihydrofolate reductase. J Am Chem Soc 133:20561–20570
132. Doshi U, McGowan LC, Ladani ST, Hamelberg D (2012) Resolving the complex role of enzyme conformational dynamics in catalytic function. Proc Natl Acad Sci USA 109(15): 5699–5704
133. Tokuriki N, Tawfik DS (2009) Protein dynamism and evolvability. Science 324:203–207

134. James LC, Tawfik DS (2003) Conformational diversity and protein evolution – a 60-year-old hypothesis revisited. Trends Biochem Sci 28:361–368
135. Zimmermann J, Oakman EL, Thorpe IF, Shi X, Abbyad P, Brooks CL 3rd, Boxer SG, Romesberg FE (2006) Antibody evolution constrains conformational heterogeneity by tailoring protein dynamics. Proc Natl Acad Sci USA 103:13722–13727
136. Khersonsky O, Tawfik DS (2010) Enzyme promiscuity: a mechanistic and evolutionary perspective. Annu Rev Biochem 79:471–505
137. Hou L, Honaker MT, Shireman LM, Balogh LM, Roberts AG, Ng KC, Nath A, Atkins WM (2007) Functional promiscuity correlates with conformational heterogeneity in A-class glutathione S-transferases. J Biol Chem 282:23264–23274
138. O'Brien PJ, Herschlag D (1999) Catalytic promiscuity and the evolution of new enzymatic activities. Chem Biol 6:R91–R105
139. Jimenez R, Salazar G, Yin J, Joo T, Romesberg FE (2004) Protein dynamics and the immunological evolution of molecular recognition. Proc Natl Acad Sci USA 101:3803–3808
140. James LC, Tawfik DS (2009) The specificity of cross-reactivity: promiscuous antibody binding involves specific hydrogen bonds rather than nonspecific hydrophobic stickiness. Protein Sci 12:2183–2193
141. Lin Y-S, Hsu W-L, Hwang J-K, Li W-H (2007) Proportion of solvent-exposed amino acids in a protein and rate of protein evolution. Mol Biol Evol 24:1005–1011
142. Jernigan RL, Kloczkowski A (2007) Packing regularities in biological structures relate to their dynamics. Methods Mol Biol 350:251–276
143. Brown CJ, Takayama S, Campen AM, Vise P, Marshall TW, Oldfield CJ, Williams CJ, Dunker AK (2002) Evolutionary rate heterogeneity in proteins with long disordered regions. J Mol Evol 55:104–110
144. Woycechowsky KJ, Choutko A, Vamvaca K, Hilvert D (2008) Relative tolerance of an enzymatic molten globule and its thermostable counterpart to point mutation. Biochemistry 47:13489–13496
145. Gould SM, Tawfik DS (2005) Directed evolution of the promiscuous esterase activity of carbonic anhydrase II. Biochemistry 44:5444–5452
146. Umeno D, Tobias AV, Arnold FH (2005) Diversifying carotenoid biosynthetic pathways by directed evolution. Microbiol Mol Biol Rev 69:51–78
147. Brustad EM, Arnold FH (2011) Optimizing non-natural protein function with directed evolution. Curr Opin Chem Biol 15:201–210
148. Williams GJ, Zhang C, Thorson JS (2007) Expanding the promiscuity of a natural-product glycosyltransferase by directed evolution. Nat Chem Biol 3:657–662
149. Jourden MJ, Clarke CN, Palmer AK, Barth EJ, Prada RC, Hale RN, Fraga D, Snider MJ, Edmiston PL (2007) Changing the substrate specificity of creatine kinase from creatine to glycocyamine: evidence for a highly evolved active site. Biochim Biophys Acta 1774: 1519–1527
150. Chen ZL, Katzenellenbogen BS, Katzenellenbogen JA, Zhao HM (2004) Directed evolution of human estrogen receptor variants with significantly enhanced androgen specificity and affinity. J Biol Chem 279:33855–33864
151. Joerger AC, Mayer S, Fersht AR (2003) Mimicking natural evolution in vitro: an N-acetylneuraminate lyase mutant with an increased dihydrodipicolinate synthase activity. Proc Natl Acad Sci USA 100:5694–5699
152. Gaille C, Kast P, Haas D (2002) Salicylate biosynthesis in *Pseudomonas aeruginosa*. Purification and characterization of PchB, a novel bifunctional enzyme displaying isochorismate pyruvate-lyase and chorismate mutase activities. J Biol Chem 277:21768–21775
153. DeGrado WF (1993) Peptide engineering. Catalytic molten globules. Nature 365:488–489
154. Guarnera E, Pellarin R, Caflisch A (2009) How does a simplified-sequence protein fold? Biophys J 97:1737–1746
155. Suskiewicz MJ, Sussman JL, Silman I, Shaul Y (2011) Context-dependent resistance to proteolysis of intrinsically disordered proteins. Protein Sci 20:1285–1297

Top Curr Chem (2013) 337: 69–94
DOI: 10.1007/128_2012_418

Published online: 12 March 2013

A Surprising Role for Conformational Entropy in Protein Function

A. Joshua Wand, Veronica R. Moorman, and Kyle W. Harpole

Abstract Formation of high-affinity complexes is critical for the majority of enzymatic reactions involving proteins. The creation of the family of Michaelis and other intermediate complexes during catalysis clearly involves a complicated manifold of interactions that are diverse and complex. Indeed, computing the energetics of interactions between proteins and small molecule ligands using molecular structure alone remains a great challenge. One of the most difficult contributions to the free energy of protein–ligand complexes to access experimentally is that due to changes in protein conformational entropy. Fortunately, recent advances in solution nuclear magnetic resonance (NMR) relaxation methods have enabled the use of measures-of-motion between conformational states of a protein as a proxy for conformational entropy. This review briefly summarizes the experimental approaches currently employed to characterize fast internal motion in proteins, how this information is used to gain insight into conformational entropy, what has been learned, and what the future may hold for this emerging view of protein function.

Keywords Binding free energy · Conformational entropy · Drug design · Michaelis complexes · Molecular recognition · Motional proxy · NMR relaxation

Contents

A.J. Wand (✉), V.R. Moorman, and K.W. Harpole
Graduate Group in Biochemistry & Molecular Biophysics, The Johnson Research Foundation and Department of Biochemistry & Biophysics, University of Pennsylvania Perelman School of Medicine, Philadelphia, PA 19104-6059, USA
e-mail: wand@mail.med.upenn.edu

Abbreviations

CaM	Calcium-saturated calmodulin
CaMK1(p)	CaM-binding domain of the calmodulin kinase I
CaMKKα(p)	CaM-binding domain of calmodulin kinase kinase alpha
CSA	Chemical shift anisotropy
DHF	Dihydrofolate
DHFR	Dihydrofolate reductase
eNOS(p)	CaM-binding domain of the epithelial nitric oxide synthase
	GlcNAc *N*-acetyl-D-glucosamine
HEWL	Hen egg white lysozyme
J(ω)	Lipari–Szabo model-free spectral density at frequency ω
NADPH/NADP$^+$	Reduced/oxidized nicotinamide adenine dinucleotide phosphate
NMR	Nuclear magnetic resonance
nNOS(p)	CaM-binding domain of the neuronal nitric oxide synthase
	NOE Nuclear Overhauser effect
O^2	Lipari–Szabo model free squared generalized order parameter
PDE(p)	CaM-binding domain of the phosphodiesterase
smMLCK(p)	CaM-binding domain of the myosin light chain kinase
SW	Square well potential energy function
T_1 (R_1)	Longitudinal relaxation time (longitudinal relaxation rate)
T_2 (R_2)	Transverse relaxation time (transverse relaxation rate)
ΔS_i	Change in entropy for the *i*th component
τ_e	Lipari–Szabo model-free effective correlation time
τ_m	Isotropic macromolecular tumbling correlation time

1 Introduction

Molecular recognition by proteins is fundamental to almost every biological process, particularly those protein–ligand complexes underlying enzymatic catalysis. The process of molecular recognition has therefore been of central interest from the earliest days of enzymology. Various mechanistic frameworks have been constructed to describe the physical basis for specific high affinity interactions between protein molecules and their ligands. They have ranged from the simple "lock and key" interaction model [1] to the "induced fit" [2] and the more recent "conformational selection" models [3, 4]. Each of these derives from an increasing recognition of the plasticity and energetics of the ensemble of structures that proteins occupy [5, 6] and the potential role for travelling across this complex energy landscape in protein function [7, 8].

For the most part, the predominant interest has been in characterizing how motion between structural states correlate with protein function. There have been spectacular demonstrations of conformational selection (e.g., [9, 10]) and strong indications for the participation of "special" protein motions in catalysis by enzymes (e.g., [11–14]), though the latter role has been vigorously criticized on general principles [15] (but see [16]). What concerns us here is not that proteins move from one functional state to another but rather what is the thermodynamic basis of these functionally relevant states. This is a long-standing issue and is intimately related to the so-called protein folding problem and the free energy landscape that protein molecules explore [17]. Stated less obliquely, we wish to understand the physical basis for the (de)stabilization of a given state of a protein molecule. Perhaps the most primitive and potentially simplest function that a protein engages in is high affinity molecular recognition. The formation of protein complexes involves a complicated manifold of interactions that are diverse and complex. This complexity is reflected in the difficulty of computing the energetics of interactions involving proteins using molecular structure alone [18–20]. Indeed, structure-based design of pharmaceuticals has been impeded by this barrier [21]. Here, we specifically focus on the role of protein conformational entropy in modulating the free energy of the association of a protein with a ligand.

Expression of the total binding free energy emphasizes that the entropy of binding is comprised of contributions from the protein, the ligand and the solvent:

$$\Delta G_{\text{bind}} = \Delta H_{\text{bind}} - T(\Delta S_{\text{protein}} + \Delta S_{\text{ligand}} + \Delta S_{\text{solvent}}) \qquad (1)$$

The free energy of binding (ΔG_{bind}) and the enthalpy of binding (ΔH_{bind}) can in favorable cases be directly measured using isothermal titration calorimetry [22] and the associated total binding entropy (ΔS_{bind}) obtained by arithmetic. Unfortunately, as (1) indicates, the molecular origins of these thermodynamic parameters remain obscure. Views of associations involving proteins have largely been seen through the lens of enthalpy owing to the richness of our knowledge of the structure of proteins and their complexes, which helps reveal the details of the interactions governing the enthalpy. In great contrast, the origin of the change in entropy remains difficult to grasp as it inherently involves a manifold of states that the protein, ligand, and solvent can occupy, each having its own probability for existence. Historically, the contributions by solvent entropy to binding thermodynamics have taken center stage and are usually framed in terms of the hydrophobic effect [23]. Hydrophobic solvation by water continues to be the subject of extensive analysis [24, 25].

In principle, the entropic contribution of a structured protein to the binding of a ligand ($\Delta S_{\text{protein}}$) includes both changes in its internal conformational (configurational) entropy (ΔS_{conf}) and changes in rotational and translational entropy (ΔS_{RT}) [26]. Equation (1) emphasizes that the measurement of the entropy of binding does not resolve contributions from internal protein conformational entropy. Arguments from fundamental theory [27] and observations from simulation (e.g., [28, 29]) and experiment (e.g., [30, 31]) in the 1970s indicated that proteins fluctuate about a structure closely similar to that observed by crystallography and that these fluctuations could reflect significant residual conformational entropy. Yet it is only recently that experimental methods and strategies have been created to assess this and related ideas quantitatively.

Experimental measurement of ΔS_{conf} has been difficult. During the past decade, we and others have been developing NMR methods that serve to provide measures of motion between different microscopic structural states and thus can act as an indirect measure of or proxy for conformational entropy [32]. The idea is simple, but extremely tricky to implement. However, as outlined below, solution NMR spectroscopy has emerged as the most powerful experimental technique for accessing protein motion in a site-resolved comprehensive manner. Motion expressed on the sub-nanosecond time scale corresponds to significant entropy [33, 34] and NMR relaxation methods are particularly well suited for its characterization [32]. As will be illustrated in detail below, this leads directly to the idea of using measures of motion as a proxy for conformational entropy. Thus, detection of motion segues into the issue of conformational entropy, though it is not obvious how to employ measures of motion as a quantitative measure of conformational entropy. Of course, one must first enable the site resolved measurement of internal motion (disorder) of proteins.

2 Solution NMR Spectroscopy and Detection of Motion

Over the past two decades, solution NMR spectroscopy has emerged as a powerful means for the site-resolved measurement of motion on an impressive range of time scales in proteins of significant size [32, 35–37]. The breadth of available time scales is daunting and leaves one asking where, with limited resources, one should begin. Here we are ultimately most focused on those motions that *express* large contributions to protein conformational entropy. Simple arguments suggest that this should be largely manifested in the extremely fast motions corresponding to bond vibrations and torsional oscillations, which generally occur on the nanosecond and faster time scales [33]. Classical NMR relaxation phenomena allow access to this time scale.

Very briefly, an ensemble of nuclear spins can relax from a non-equilibrium state via a range of potential interactions. For example, the inter-nuclear dipole–dipole interaction between spatially proximal hydrogens gives rise to the nuclear Overhauser effect (NOE), which is likely familiar to the non-expert reader as a means to measure inter-atomic distances for the determination of molecular structure in solution. The strength of this interaction depends on the time average of the distance between the nuclei and of the angle of the inter-nuclear vector with the applied magnetic field. In favorable situations, such as when the two nuclei are bonded, the distance dependence is effectively constant (though see [38, 39]) and the interaction is temporally modified only by the change in the orientation of the bond vector with respect to the magnetic field. Examples include the $^{15}N–^{1}H$ amide bond and the $^{13}C–^{1}H$ bond in a variety of contexts such as in a methyl group. Other nuclear spin interactions are perhaps more obscure to the reader and include chemical shift anisotropy (CSA), which reflects the interaction of the nuclear spin with the asymmetric distribution of electrons about it, and the electrostatic interaction between the quadrupolar moment of a nucleus having a spin quantum number of one or greater and the surrounding electric field gradient. In some cases, the "interference" or "cross correlation" between different relaxation mechanisms can offer insight into motion (e.g., [40–45]), though these approaches are not common owing to a variety of technical limitations.

Relaxation of nuclear spins in liquid samples to an equilibrium distribution of spin states is mediated by the fluctuation of local fields. Rapid molecular motions impose time modulation on these local fields and it is through this dependence that information about motion can be obtained. The theoretical treatment of the connection between motion and NMR relaxation phenomena is complicated. The interested reader is referred to a recent monograph on this and related subjects [46]. A popular and very robust way of capturing the essential character of the motion that gives rise to NMR relaxation phenomena of the type considered here is the so-called model free approach of Lipari and Szabo [47, 48]. The various NMR observables that can be measured can be generally expressed as linear combinations of the so-called spectral density functions. The spectral densities are in turn defined by the motion of the "interaction" vector within the protein. Consider a ^{13}C nucleus attached by a single ^{1}H, i.e., in a CHD_2 isotopomer. The deuterium nucleus has a much weaker interaction than the ^{1}H nucleus and its effects can be largely ignored (though they are not in the final analysis). The motion of the $^{13}C–^{1}H$ interaction vector (in this case along the rigid bond between them) can be described by an autocorrelation function, which is simply the dot product of the interaction vector's orientation at some time t and its orientation at some time t'. The time dependence will have two components: a contribution from the slower global tumbling of the protein and an assumed faster component due to motion within the molecular frame. In the simplest case, the Lipari–Szabo treatment leads to three parameters: a correlation time for isotropic macromolecular reorientation (τ_m), an effective correlation time (τ_e), and a measure of the angular disorder of the interaction vector termed the squared generalized order parameter (O^2).[1] The order parameter by definition ranges from zero to one, corresponding to complete isotropic disorder

[1] Please note: to avoid confusion with entropy we will refer to the Lipari–Szabo squared generalized order parameter as O^2 rather than its original designation as S^2.

and complete rigidity of the interaction vector within the molecular frame, respectively. It is this motional parameter that offers the potential to provide access to statements about conformational entropy. The effective correlation time has a strong technical definition that precludes its general use as faithful descriptor of the time constant(s) for the underlying motion.

The Lipari–Szabo model-free spectral density has a very simple form:

$$J(\omega) = \frac{O^2\tau_{\mathrm{m}}}{1+\omega^2\tau_{\mathrm{m}}^2} + \frac{(1-O^2)\tau}{1+\omega^2\tau^2} \tag{2}$$

where $\tau^{-1} = \tau_{\mathrm{m}}^{-1} + \tau_{\mathrm{e}}^{-1}$. The spectral densities are linearly combined as required by the physics of the specific NMR relaxation mechanism to define an observable relaxation (see [32] for illustrative derivations). For example, the longitudinal relaxation rate ($1/T_1$ or R_1) of the ^{13}C nucleus by a single bonded ^{1}H nucleus is given by:

$$R_1 = \frac{d^2}{4}[J(\omega_{\mathrm{H}} - \omega_{\mathrm{C}}) + 3J(\omega_{\mathrm{C}}) + 6J(\omega_{\mathrm{H}} + \omega_{\mathrm{C}})] \tag{3}$$

where d^2 is comprised of fundamental constants and the effective C–H bond length and ω_{H} and ω_{C} are the resonance frequencies of ^{1}H and ^{13}C nuclear spins, respectively. For each site of interest there are two unknowns (O^2 and τ_{e}) plus one global variable (τ_{m}) defining isotropic tumbling of the protein in solution. The situation can become more complicated, of course. For example, the character of the tumbling of the macromolecule may be anisotropic to some degree. This is easily handled using appropriate diffusion equations and data filtering [49–52].

Similarly, specific instances may require (justify) more or less complex forms of the model-free spectral density shown above [53]. Nevertheless, the experimental prescription is clear: resolve relaxation data at n individual sites in a protein and measure as many relaxation parameters (e.g., T_1, T_2, etc.) at as many magnetic fields (to vary ω_{H}, ω_{C}) as needed to provide a robust determination of the $2n + 1$ (in the case of isotropic macromolecular tumbling) parameters.

Several key steps were required to enable the comprehensive use of NMR relaxation phenomena to characterize the internal motion of proteins of significant size. The advent of multidimensional NMR spectroscopy provides a means to resolve literally hundreds of probe sites for motion in proteins. Sophisticated isotopic labeling schemes have been introduced to simplify the complexity of the NMR relaxation as much as possible in order to make its measurement and subsequent analysis more robust (e.g., [54–60]). Generally, the strategy is to reduce the number of relaxation mechanisms (interactions) as much as possible. Finally, the cornerstone was the development of NMR experiments that prepared "pure" NMR observables of NMR relaxation that could be directly interpreted. Lewis Kay and colleagues are largely identified with the development of the NMR machinery necessary to measure ^{15}N [61, 62], ^{2}H [63, 64], and ^{13}C [65] autorelaxation in proteins. More recently, Tugarinov and coworkers have extended the number of

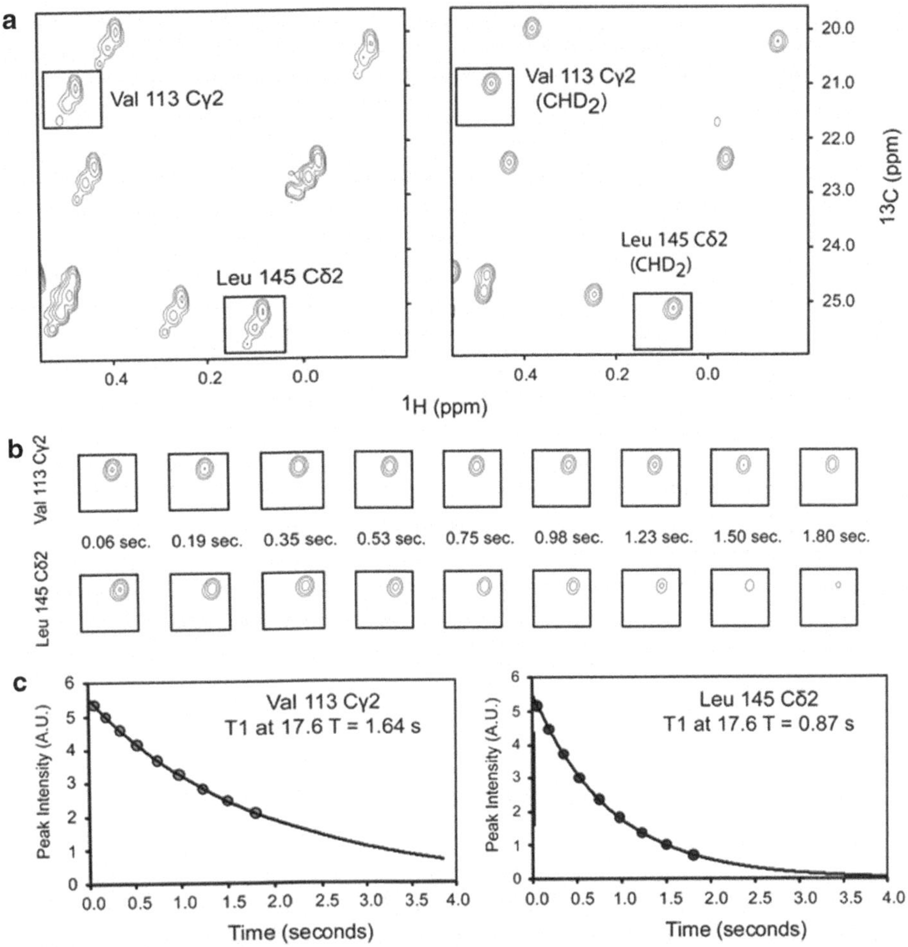

Fig. 1 Observation of carbon relaxation in proteins. (**a**) Protein is expressed during growth on $^{13}CHD_2COOD$ pyruvate in D_2O. Most methyl groups are selectively labeled with ^{13}C [57] as a mixture of deuterium isotopomers [67]. (**b**) The appropriate $^{13}CHD_2$ methyl isotopomer is selected during preparation of magnetization which (**c**) allows longitudinal and transverse relaxation to be measured in an otherwise deuterium background [65]. See [32] for further details

deuterium relaxation experiments that provide a context for extracting even more fundamental relaxation rate constants at a single magnetic field [66]. Notable contributions for implementing ^{13}C relaxation in the context of proteins came from Torchia and colleagues in their unraveling of the complexity of this particular mechanism of relaxation [67]. Finally, computational strategies were needed to extract confidently the desired model-free parameters [53, 68, 69].

The basic experimental strategy is outlined for methyl carbon relaxation in Fig. 1. To make this particular situation as simple as possible, the interaction of the methyl carbon is restricted to a single bonded ^1H. This is arranged by expression of the protein of interest during growth on randomly partially deuterated $^{13}C_3$-pyruvate as a general carbon source [57] or with unlabeled glucose and appropriately labeled metabolic precursors for valine, leucine, and isoleucine [59]. Protein expression is carried out in "100%" D_2O to ensure elimination of ^1H spins at non-methyl sites in the protein [67].

Proteins produced in this manner are largely deuterated with selective ^1H and ^{13}C labeling in methyl groups. The methyl groups are mixtures of isotopomers (i.e., CD_3, CHD_2, CH_2D, CH_3). Only the three carrying at least one hydrogen are observed in the two-dimensional ^1H–^{13}C chemical shift correlation spectrum (Fig. 1). The three observed isotopomers give crosspeaks that are at slightly different positions in the spectrum. The appropriate isotopomer, in this case the CHD_2, is selected by spectroscopic manipulation [65]. The NMR relaxation experiment is designed to follow the return of a particular type of magnetization from a non-equilibrium state back to equilibrium. For carbon relaxation in proteins, longitudinal and transverse relaxation processes are most useful and are likely familiar to the non-specialist reader as "T_1" and "T_2" relaxation, respectively. The NOE is not so useful in the context of carbon relaxation in proteins of significant size since this observable reaches a limit as molecular tumbling slows. The relaxation process at a given methyl is quantified by the variation of the intensity of the corresponding ^1H–^{13}C cross peak with the relaxation time period (Fig. 1). The variation of the rate of relaxation across sites in the protein reveals corresponding differences in the underlying dynamics.

3 Fast Motion in Proteins Observed by NMR Relaxation

For technical reasons alluded to above, the primary probes of fast ps–ns motion in proteins have been the amide N–H bond and the C–H bond in methyl groups. Here "fast" is defined by processes occurring on time scales significantly shorter than the macromolecular tumbling time of the protein in solution, which is generally on the order of 50 ns or less in the context of current studies. The ^{15}N experiments are relatively straightforward and there are literally hundreds of studies of proteins spanning a significant range of topologies and contexts [36]. The view of the backbone provided by amide ^{15}N relaxation is generally unimpressive with uniformly high order parameters (rigidity) in regions of large elements of regular secondary structure (i.e., beta sheets, helices), which are bounded by the termini and intervening loops, turns, and sections of irregular structure having significantly larger amplitude motion [36]. Changes in functional state generally have small and often quite subtle effects on the backbone of the protein. Overall, the polypeptide chain appears largely to be a relatively rigid scaffold. Most of the dynamic response in the ps-ns time regime to changes in functional state is seen to reside in the side chains [32]. Observation of methine and methylene carbon hydrogens is complicated by the difficulty of appropriate labeling [55, 56, 70]. For probing the motion of methyl-bearing amino acid side chains, deuterium relaxation is preferred due to

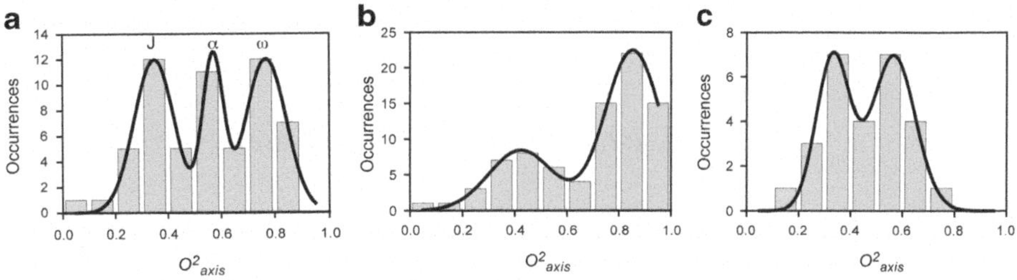

Fig. 2 Histograms of the distribution of squared generalized order parameters of methyl group symmetry axes (O^2_{axis}) in (**a**) the complex of calcium-saturated calmodulin and a peptide derived from the calmodulin-binding domain of the smooth muscle myosin light chain kinase [72], (**b**) flavodoxin [73], and (**c**) α_3D, a protein of de novo design [74]. *Lines* are best fits to a sum of three Gaussians. Taken from Igumenova et al. [32]. Copyright American Chemical Society

the purity of the source of its relaxation [32]. However, again for technical reasons, this approach is largely restricted to proteins smaller than ~25 kDa due to the effects of slow macromolecular reorientation in solution. For larger proteins, carbon relaxation methods prove to be the most robust though preparation of the samples and analysis of the data is somewhat more involved [67, 71]. The motion of methyl-bearing side chains in several dozen proteins have been studied in comprehensive detail using these methods [32]. Though sometimes obscured due to limited sampling in small proteins, three types or classes of motion of methyl-bearing side chains have been revealed by the distribution of methyl symmetry axis Lipari–Szabo order parameters [32]. The so-called J-class is centered around an O^2_{axis} value of ~0.35 and involves motion of the methyl group between rotameric wells, leading to averaging of the associated J-coupling. The α-class is centered around an O^2_{axis} value of ~0.65 and has smaller contribution from motions that lead to rotameric interconversion and generally reflects large amplitude motion within a single rotameric well. The *ω*-class is centered around an O^2_{axis} value of ~0.85 and has highly restricted motion within a single rotameric well that is somewhat reminiscent of the uniform rigidity of most backbone sites.

What is fascinating about these classes of motion is that their distribution in individual proteins can be quite variable (Fig. 2). The significant variation in the effective amplitude of motion through the protein matrix is perhaps somewhat counterintuitive. For example, restriction of motion is not strongly correlated with the depth of burial and other simple structural features [32]. Clearly, the "rules" for motion in proteins remain to be fully understood.

4 Motional Proxy for Conformational Entropy

Some time ago Akke and coworkers [75] introduced a theoretical scheme that provided a parametric connection between motion captured by NMR relaxation and the thermodynamic parameters of an ensemble of motional probes. This is directly revealed by the formal definition of the Lipari–Szabo squared generalized order parameter [47]:

$$O^2 = \iint p_{eq}(\Omega_1)P_2(\cos\theta_{12})p_{eq}(\Omega_2)\mathrm{d}\Omega_1\,\mathrm{d}\Omega_2 \tag{4}$$

where Ω_1 and Ω_2 represent separate orientations (states) of the NMR interaction vector, p_{eq} is the corresponding probability of each state, and P_2 the second order Legendre polynomial of the cosine of the angle θ_{12} between the two states. Clearly, the explicit consideration of the probability of the various states accessible to the NMR relaxation probe provides a direct connection to the partition function governing the ensemble. Hence, in principle, one can have access to the fundamental thermodynamic parameters of the ensemble through the usual relations. However, this approach requires a specific model (potential energy) for the motion in order to make the parametric connection between what can be measured (O^2) and what is desired (S).

Adopting this idea, we [34] and Yang and Kay [76] used the simple harmonic oscillator and diffusion in an infinite square well potential, respectively, to illustrate the parametric relationship between the Lipari–Szabo squared generalized order parameter (O^2) and entropy (S). The uses of motion as a proxy for or as an indirect measure of conformational entropy and some of the fundamental issues associated with this strategy are illustrated in Fig. 3. Consider an amino acid side chain with a single degree of motional freedom such as a point about which the side chain can pivot. Three different simple potentials based on the angle of the pivot are illustrated in Fig. 3. The square well potential has an infinite barrier set at some fluctuation angle. This potential corresponds to free diffusion in a cone, which has an analytical expression relating the Lipari–Szabo order parameter to the corresponding entropy [76]. This infinite square well potential is somewhat unrealistic as it does not express thermodynamic some properties such as heat capacity, which are an essential descriptor of protein molecules [77].

Also illustrated in Fig. 3 are the classic quadratic harmonic oscillator and its stiffer sixth power colleague, which do have the ability to represent a broader range of thermodynamic attributes of proteins in this context [75, 78]. The family of potentials of Fig. 3 gives entropies that are off-set from each other. Given the persistent uncertainty of the precise nature of the potential governing protein motion, the determination of *absolute* entropy is generally not possible. Fortunately, as Fig. 3 also indicates, the dependence of the slope (i.e., $\mathrm{d}S/\mathrm{d}O^2$) is relatively insensitive to the nature of the underlying potential. Thus *differences* in entropy obtained from *differences* in measures of motion seems to be possible [34]. Thus the "dynamical proxy" for entropy would appear to be useful for analysis of changes in motion corresponding to variation of the Lipari–Szabo order parameter between ~0.1 and ~0.9 even in the presence of a possible change in the underlying potential energy function upon a change in state (e.g., a binding event, pressure and temperature change, etc.).

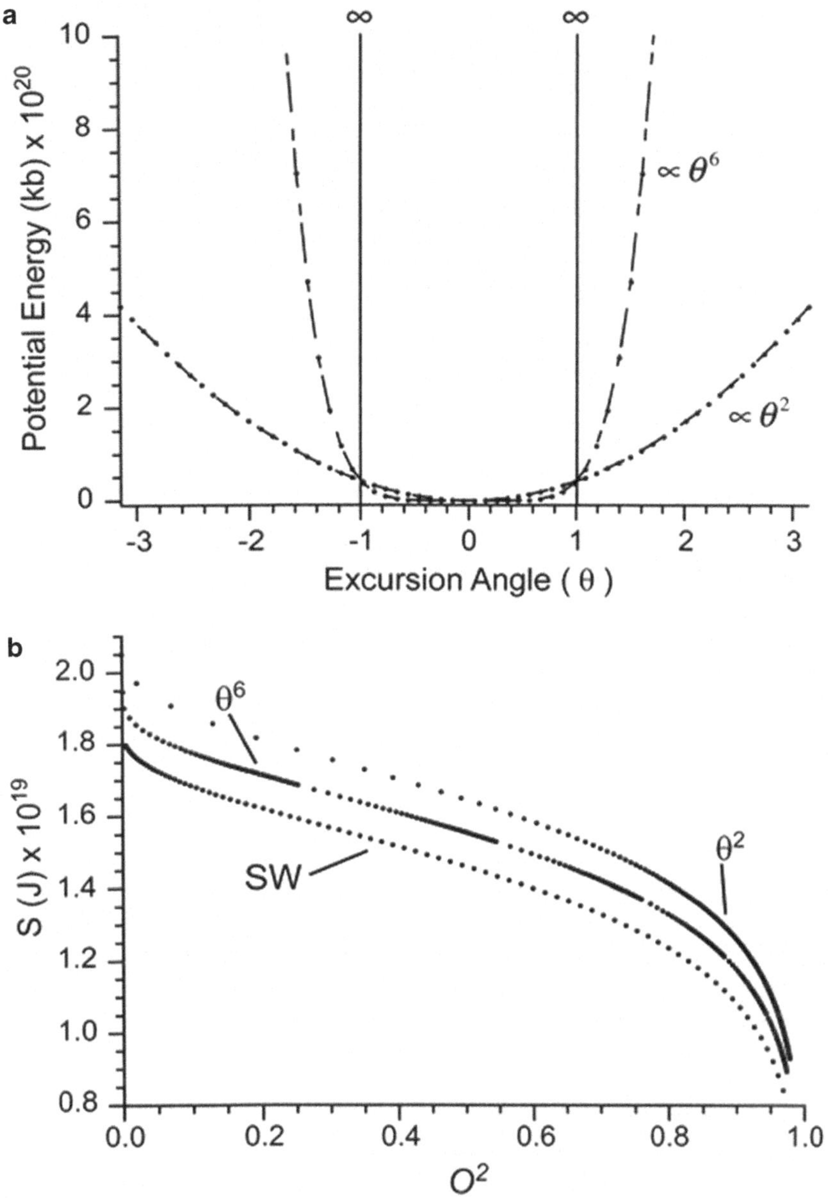

Fig. 3 The dynamical proxy for entropy. (**a**) Various simple potential energy functions governing the behavior of an attached NMR relaxation "spy". Shown are the infinite square well potential (*SW*) with an angular barrier of 1 radian, the simple harmonic oscillator potential with a quadratic dependence on the excursion angle (θ^2) and its sixth-power cousin (θ^6). (**b**) Parametric relationships between the various potentials and the corresponding Lipari–Szabo squared generalized order parameter. See [34, 75, 76, 78] for further explanation and examples

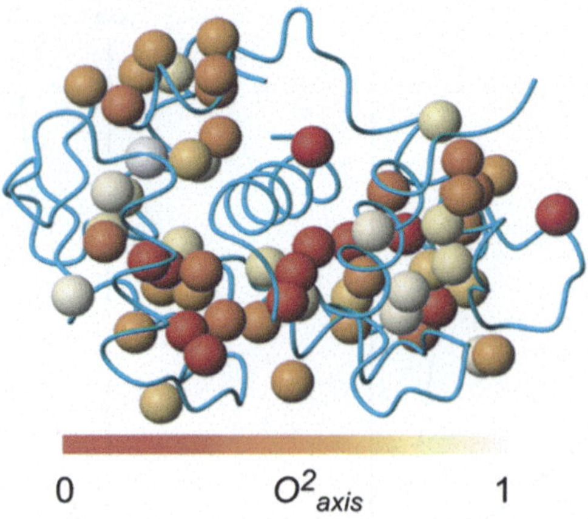

Fig. 4 *Ribbon* representation of the complex of calcium-saturated calmodulin and a peptide corresponding to the calmodulin-binding domain of the calmodulin kinase I. Methyl groups of calmodulin are represented as *spheres* and are shaded according to Lipari–Szabo order parameter of the methyl symmetry axis (O^2_{axis}) as determined by deuterium relaxation methods. Taken from Frederick et al. [91]. Copyright American Chemical Society

In an approach that we term an "oscillator inventory," a simple prescription is used to estimate changes in conformational entropy from changes in protein internal motion: (1) measure as many differences in Lipari–Szabo order parameters between two states of the protein as possible; (2) look-up the corresponding changes in entropy (i.e., using the type of relationship shown in Fig. 3); and (3) take the differences and simply add them up. There are many legitimate objections to this approach, including concerns about correlated motion, incomplete sampling, validity of the potential, the dependence of the observed relaxation on the geometric details of the motion, etc., all of which can potentially compromise the interpretation [32].

Despite the obvious limitations of the "oscillator inventory" approach, it has led to observations that promote the general idea that ligand binding to proteins produces significant changes in internal motion that, in turn, correspond to significant changes in conformational entropy (e.g., [72, 79–85]). We have used calmodulin and its multitude of binding partners as a model system to investigate the role of conformational entropy in high affinity association of proteins [86]. Calmodulin is central to calcium-mediated signal transduction pathways of eukaryotes [87] and interacts with hundreds of proteins with high affinity [88]. The structures of the calmodulin complexes generally follow a "hot dog in a bun" type of topology where the two globular domains of calmodulin collapse around the target calmodulin-binding domain, which forms an amphiphilic helix that is largely sequestered from solvent. Calmodulin-binding domains are generally 20–30 residue sequences characterized by enrichment in basic and hydrophobic residues [88]. This system is particularly amenable to deuterium methyl relaxation experiments [32, 72, 78, 86, 89, 90]. As shown in Fig. 4, methyl probes are distributed throughout the calmodulin molecule. The target domains also have strong representation by the methyl-bearing amino acids [89].

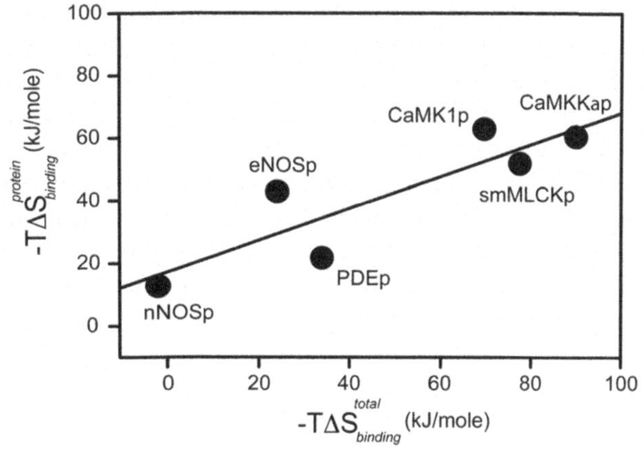

Fig. 5 Application of the oscillator inventory approach to the calmodulin complexes. CaM complexes with six natural target domains were employed. The change in conformational entropy was estimated from the changes in methyl group symmetry axis squared generalized order parameters measured by deuterium relaxation [86]. The simple harmonic oscillator model was used. The linear correlation coefficient (R^2) of conformational entropy vs the entropy of binding is 0.78. Adapted from Frederick et al. [86] with permission. Copyright Nature Publishing

Using NMR relaxation measurements, we have shown that calcium-saturated calmodulin (CaM) is an unusually dynamic protein that is characterized by a broad range of the amplitudes (i.e., order parameters) of fast side-chain dynamics that are redistributed upon binding of a CaM-binding domain of a regulated protein [72]. Particularly intriguing is the observation that the conformational entropy estimated by the "oscillator inventory" approach correlated linearly with the overall entropy of binding of a series of CaM-binding domains to CaM (Fig. 5) [86]. It must be emphasized that there is no physical law that requires a linear correlation of a single component of the total entropy (i.e., the conformational entropy) with the total entropy. However, the persistence of such a linear correlation suggests a biological origin: Nature employs conformational entropy in evolving toward the optimal free energy of binding. To first approximation this makes sense if a primary selection pressure for evolution of a protein–ligand interaction is the free energy of association then all sources of entropy will be invoked, to the extent possible, to satisfy this evolutionary demand.

5 Creation and Calibration of an "Entropy Meter"

A more recent approach is perhaps less assailable and more convincing than the simple "oscillator inventory." The idea is to subsume the various microscopic concerns enumerated above and find an *empirical* calibration. Thus measures of motion are used in a largely model-independent way, which thereby circumvents the microscopic details that are difficult to accommodate in a model-dependent calculation. In a sense, we have simply created an "entropy meter," analogous to a thermometer. It is important to note that in this approach the motion of the methyl group is used to sense its local surroundings. This relies on the coupling of motion within the protein such that the "probe" methyl groups report on the local disorder [90]. This is a critical change in perspective where the dynamical probe is interpreted to reflect not only its own disorder but also the disorder of the surrounding non-probe protein matrix. Thus, for the approach to work there must be sufficient coupling between the motion of the probe and its surroundings and there must be a sufficient density of probes to provide adequate coverage of the protein.

5.1 Derivation of the Dynamical Proxy "Entropy Meter"

Making several simple assumptions regarding the nature of the free states so that the solvation entropy can be calculated, one can obtain in this case [90]

$$\Delta S_{\mathrm{conf}} = m\left[\left(n_{\mathrm{res}}^{\mathrm{CaM}} \bullet \left\langle \Delta O_{\mathrm{axis}}^{2} \right\rangle^{\mathrm{CaM}} + n_{\mathrm{res}}^{\mathrm{target}} \bullet \left\langle \Delta O_{\mathrm{axis}}^{2} \right\rangle^{\mathrm{target}}\right)\right] + \Delta S_{\mathrm{other}} \tag{5}$$

$$(\Delta S_{\mathrm{tot}} - \Delta S_{\mathrm{sol}}) = m\left[\left(n_{\mathrm{res}}^{\mathrm{CaM}} \bullet \left\langle \Delta O_{\mathrm{axis}}^{2} \right\rangle^{\mathrm{CaM}} + n_{\mathrm{res}}^{\mathrm{target}} \bullet \left\langle \Delta O_{\mathrm{axis}}^{2} \right\rangle^{\mathrm{target}}\right)\right] + \Delta S_{\mathrm{RT}} + \Delta S_{\mathrm{other}} \tag{6}$$

where ΔS_{tot}, ΔS_{sol}, ΔS_{conf}, ΔS_{RT}, and $\Delta S_{\mathrm{other}}$ are the changes in total system entropy, solvent entropy, conformational entropy, rotational-translational entropy, and undocumented entropy, which is mostly solvent entropy from ion pair dissociation and solvation. The changes in side chain motion are assessed from changes in methyl order parameters weighted by the number of residues in calmodulin ($n_{\mathrm{res}}^{\mathrm{CaM}}$) and the target domains ($n_{\mathrm{res}}^{\mathrm{target}}$). By postulate, ΔS_{conf} is linearly related to the residue-weighted change in the dynamics of the target domain and the protein (e.g., calmodulin) upon binding. Linearity is strongly supported by the simple simulations illustrated in Fig. 3 [78]; "m" is the desired empirical scaling factor relating changes in motion to changes in conformation entropy.

ΔS_{tot} is obtained from isothermal titration calorimetry. ΔS_{sol} is obtained from the change in accessible surface area revealed by the structures of free CaM and the complexes. Despite some limitations of this particular system (see below), the approach worked very well. Figure 6 shows the empirical calibration of the "entropy meter" using five calmodulin complexes. The five complexes give an excellent linear relationship ($R = 0.95$) and a slope (m) of -0.037 ± 0.007 kJ K^{-1} mol res^{-1}.

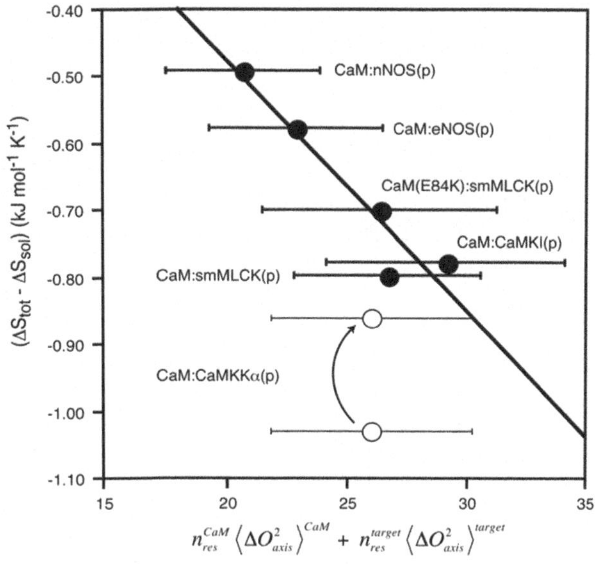

Fig. 6 Calibration of the dynamical proxy for protein conformational entropy. Simple considerations lead to the prediction of a quantitative linear relationship between the total binding entropy and the entropy of solvent to the conformational entropy by NMR relaxation parameters derived from methyl bearing amino acids (see (6)). The dynamics of free CaM and six CaM complexes were determined by deuterium methyl relaxation [89]. The lower CaM:CaMKKα(p) datum is a clear outlier, likely due to residual structure in the free CaMKKα(p) peptide. The upper CaM:CaMKKα(p) point results from a simple correction. Excluding the CaM:CaMKKα(p) complex gives the *regression line* shown. The slope of -0.037 ± 0.007 kJ K^{-1} mol res^{-1} allows for empirical calibration of the conversion of changes in side-chain dynamics to a quantitative estimate of changes in conformational entropy. The ordinate intercept is 0.26 ± 0.18 kJ K^{-1} mol res^{-1}. Reproduced with permission from Marlow et al. [89]. Copyright Nature Publishing

5.2 *Quantitative Evaluation of Entropy in Molecular Recognition by Calmodulin*

The calibration of the "entropy meter" allows the contribution of conformational entropy to the binding entropy to be evaluated quantitatively (Fig. 7). Scaling of observed changes in dynamics in calmodulin to real entropy units provides, for the first time, a quantitative statement of how changes in the conformational entropy of a protein contribute significantly to the overall binding free energy of a protein–ligand complex [90]. The basic result is striking: the conformational entropy of calmodulin is a significant component of the free energy of binding of the target domains. Indeed, the variation of the conformational entropy of calmodulin effectively "tunes" the binding entropy [89].

It is important to point out that this first example of empirical calibration of a dynamical "entropy meter" used a less than ideal system. First, access to the solvent entropy of binding relied on the assumption that the free target domains

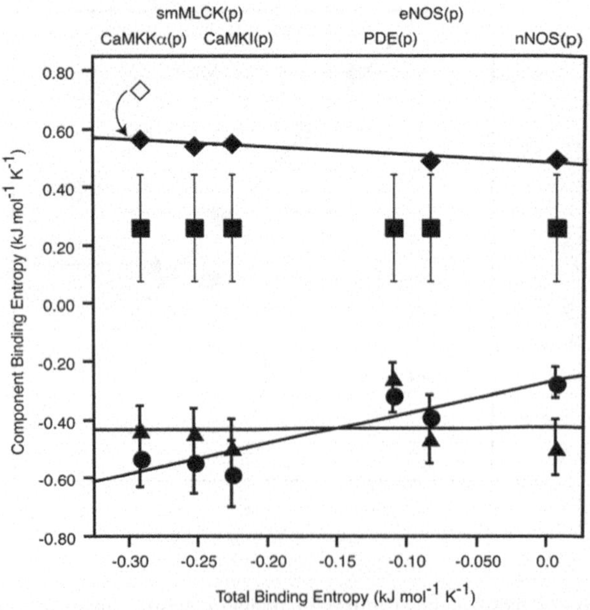

Fig. 7 Decomposition of the entropy of binding of target domains to calcium-saturated calmodulin. Based on (6) and the calibration of the dynamical proxy (see Fig. 6). *Solid diamonds* are the solvent entropies calculated from the changes in accessible surface area and include the correction resulting from the postulated hydrophobic cluster of the free CaMKα(p) target domain. The uncorrected value is shown as an *open diamond*. No structure is available for the CaM:PDE(p) complex so its solvent entropy cannot be calculated. *Solid circles* and *triangles* are the contributions to the binding entropy by the conformational entropy of CaM and the target domains, respectively. *Solid squares* are the contributions to the binding entropy not reflected in the measured dynamics (see (6)), which is obtained from linear regression. Reproduced with permission from Marlow et al. [89]. Copyright Nature Publishing

(represented as peptides) are fully solvated, i.e., completely random coil. This enabled determination of the accessible surface area of the free peptide ligand. As the CaMKKα(p) example illustrated, this is not always warranted. In addition, the calculation of solvent entropy does not consider the creation (or removal) of explicit charge. This is potentially an issue with the calmodulin complexes as each contains intimate and buried ion pairs bridging the components. It is also assumed that there are no differences in the contribution of rotation and conformational entropy to the binding free energy across the various complexes. Another important assumption is that the entropy of the free ligand can be represented by the dynamics of an unhindered methyl group. Undocumented variation in either $\Delta S_{\text{solvent}}$ or ΔS_{RT} or ΔS_{conf} of the ligand would tend to scatter the points. Fortunately, variation of these and other contributions assumed to be constant across the complexes was not a factor [except in the case of the CaMKKα(p) complex]. Nevertheless, it is clear that similar efforts are most ideally carried out on a series of complexes where ambiguity in solvation and rotational and translational entropy is less worrisome. An example of the former approach is shown in Fig. 6 where a point mutant in CaM perturbs the thermodynamics of binding to provide a useful calibration point.

Because the calmodulin complexes represent the first example of the empirical dynamical entropy meter approach, it remains unknown whether or not the empirical scaling constant is generally applicable. It would be surprising if the fundamental nature of the protein molecule giving rise to the motional coupling underpinning the approach would vary dramatically. Nevertheless, modest variation would seem entirely possible and this issue awaits further exploration of additional systems.

5.3 Future Strategies

Though the first application of the empirical "entropy meter" approach employing a dynamical proxy was apparently successful, the use of the calmodulin system highlights several potential weaknesses. In the case of calmodulin complexes, the predominant strategy was to compare complexes of different ligand target domains. Though very similar in amino acid composition and size, this introduces uncertainty in the assumption that changes in solvent entropy can be adequately calculated and that rotational-translational entropy and other sources of entropy changes remain constant across the complexes. An alternative and presumably less problematic approach is illustrated by the lone mutant CaM complex examined (Fig. 6). By using mutants distant from the protein–ligand interface to perturb the overall entropy of binding (in part through perturbations of conformational entropy) one can carry out a calibration with the same ligand. Furthermore, if mutants are carefully chosen to avoid significant variation in structure then uncertainty in changes in rotational-translational and solvent entropy will be largely avoided. Finally, it is not yet clear how to combine different probe types. Though the simple averaging of different methyl bearing amino acid dynamics gives impressive results, one anticipates that a more sophisticated averaging reflecting the variation in degrees of freedom associated with a given methyl group should be employed. This type of consideration will be even more important as quite distinct dynamic probes are introduced (e.g., the large aromatic ring systems) to ensure adequate coverage.

6 Implications for Enzyme Catalysis

Notwithstanding the uncertainty regarding the universality of the empirical linear scaling between protein motion and conformational entropy, it is interesting to speculate what the impact of conformational entropy might be in the catalytic cycle of enzymes. Here we draw two examples from the literature: hen egg white lysozyme (HEWL) and its interaction with a natural inhibitor and *E. coli* dihydrofolate reductase (DHFR) and the transition from the binary complex to the ternary Michaelis complex.

6.1 Conformational Entropy and Inhibitor Binding to Lysozyme

HEWL was the first enzyme to have its three-dimensional structure determined by X-ray diffraction [92] and has since served as a paradigm for a wide-range of biochemical and biophysical studies. The double-displacement catalytic mechanism proposed initially by Koshland [93] involving a covalent intermediate with the substrate has supplanted [94] the long held Philips mechanism that centered on a long-lived oxocarbenium ion intermediate [92]. Here we examined the thermodynamics of the interaction of an inhibitory carbohydrate with HEWL. Lysozymes catalyze the hydrolysis of β-(1,4)-linkages between *N*-acetylmuramic acid and *N*-acetyl-D-glucosamine (GlcNAc) in peptidoglycans. Additionally, some lysozymes, including that from hen egg whites, can cleave between GlcNAc residues in chitodextrins such as chitin [95]. HEWL can accommodate up to six GlcNAc residues of a chitin polymer, each binding in six sub-sites along a cleft of the protein. Cleavage occurs at the linkage between the GlcNAc residues occupying the third and fourth subsites [94, 96] and thus does not readily occur with molecules consisting of only one (GlcNAc), two (chitobiose), or three (chitotriose) GlcNAc residues [97, 98]. These smaller molecules do, however, bind with reasonable affinity (K_d~10^{-4} to 10^{-6} M) [98, 99] and therefore act as natural competitive inhibitors. Using isothermal titration calorimetry, the binding of both chitobiose and chitotriose to HEWL has been found to be enthalpically driven over a wide range of temperatures [98]. The extent of the contribution from conformational entropy manifested as fast internal motion of the protein, however, is unclear, as indicated by (1). To examine this issue, we have recently carried out a comprehensive characterization of the sub-nanosecond time scale dynamics of the backbone and of the methyl-bearing side chains of HEWL in the apo state and in complex with chitotriose [100].

The fast sub-nanosecond dynamics of the backbone and the side chains were characterized using ^{15}N- and ^{2}H-methyl relaxation, respectively [100]. From the point of view of fast methyl-bearing side chain dynamics, HEWL is an unusually rigid protein in both its free and chitotriose complexed states. Of proteins that have been examined in this way, only those having high affinity cofactors (flavodoxin) or covalently attached prosthetic groups (cytochrome c and c_2) have comparable general rigidity [73, 101, 102]. There is no distinct spatial clustering of rigidity or flexibility in the molecular structure of either the free protein or its binary complex with chitotriose.

The response of the side chain motion to chitotriose binding is complex and heterogeneous, with some sites increasing the amplitude of their motion while others are decreased. Interestingly, a significant number of relatively rigid methyl-bearing side chains of the apo state effectively become completely immobile upon binding of the ligand. These residues form a contiguous grouping that spans the core of the protein including the two catalytic residues. This core of rigidification is capped by residues that are released from an effectively rigid state in the apo state to become more dynamic upon binding chitotriose (Fig. 8). This study appears to shed light on

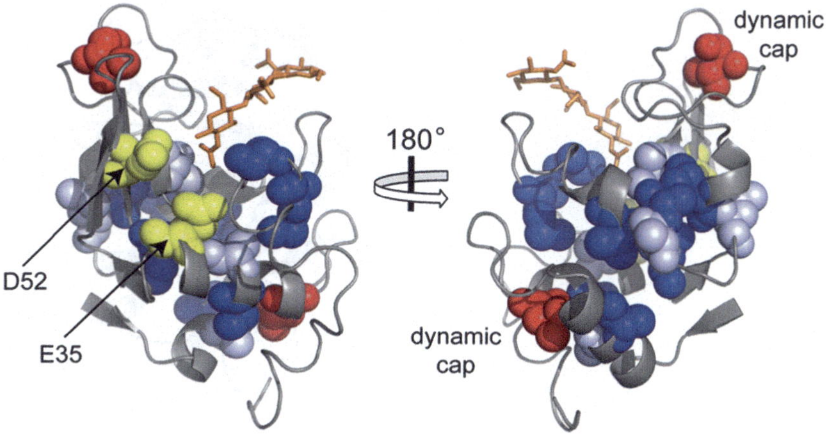

Fig. 8 Apparent cooperative rigidification of HEWL upon binding chitotriose. The backbone of the HEWL crystal structure [103] (PDB code 1LZB) is rendered as a *ribbon* and the chitotriose is shown as a *stick figure*. Atoms of residues whose methyl groups are effectively rigid in both the apo and complexed states (*light*) or become effectively rigid in the complexed state (*dark*) are shown as *spheres*. The atoms of residues whose methyl groups are effectively rigid in the apo state and become dynamic in the complex and cap the residues that are rigid in the bound state are indicated with *arrows*. The catalytic amino acids E35 and D52 are also labeled. Taken from Moorman et al. [100] with permission. Copyright Wiley-Blackwell and the Protein Society

a cooperatively formed rigidified core contacting HEWL's catalytic residues and capped by two sites that become markedly more flexible on either end. The changes in methyl axis squared generalized order parameters across the molecule average to nearly zero ($\Delta O^2_{\text{axis}} = +0.019 \pm 0.004$). If it is assumed that the empirical scaling between changes in motion and the corresponding changes in conformational entropy determined for the calmodulin complexes [89] is applicable, then by averaging over the entire lysozyme protein and scaling the resulting average change in ΔO^2_{axis} with the empirical constant of -0.037 ± 0.007 kJ mol res^{-1} K^{-1} one estimates that the response of lysozyme to binding of chitotriose corresponds to +28 ± 8 kJ mol^{-1} at 308 K.

6.2 *Conformational Entropy and Dihydrofolate Reductase*

Dihydrofolate reductase (DHFR) is a ubiquitous enzyme found in all organisms. It catalyzes the hydride transfer reaction converting dihydrofolate (DHF) to tetrahydrofolate using NADPH as its reducing cofactor. This enzyme is solely responsible for the cellular supply of THF, which serves as an essential metabolic precursor for DNA biosynthesis [104]. The catalytic cycle involves five intermediates, all of which involve at least one bound ligand. Following the hydride transfer step, the protein undergoes a conformational change in an active site loop

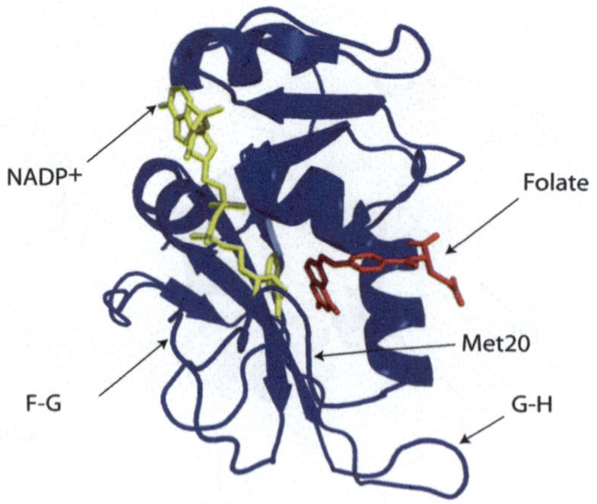

Fig. 9 *Ribbon* diagram of the *E. coli* DHFR enzyme in complex with $NADP^+$ and folate. Drawn with PyMol (Schrödinger, Portland, Oregon). Based on PDB code 1RX2 [108]

referred to as the Met 20 loop (Fig. 9). This loop moves from the closed state, in which it shields the reactants from solvent, to the occluded state, where the nicotinamide ring of NADPH is blocked from the active site. Two other proximal loops, the F–G and G–H loops, stabilize the two states of the Met 20 loop through hydrogen bonding interactions [104]. The dynamics of these loops appears to play an important role in catalysis [105]. Studies of sub-nanosecond side chain dynamics of DHFR have also been performed and show significant changes in methyl order parameters during different stages of the catalytic cycle [106]. An interesting example impacts our understanding of the role of conformational entropy during the catalytic cycle. Wright and coworkers [106] determined the Lipari–Szabo methyl symmetry axis order parameters of *E. coli* DHFR ternary complex with $NADP^+$ and folate, which is a generally accepted model for the DHFR·NADPH·dihydrofolate Michaelis complex. Lee and coworkers [107] have carried out a similar study of the DHFR·NADPH binary complex. Though done under slightly different experimental conditions (*T*, pH, buffer), the average change in $O^2{}_{\mathrm{axis}}$ of about +0.02 on going from the binary to the ternary complex combined with the scaling constant of Marlow et al. [89] suggests that an unfavorable reduction in entropy of ~0.1 kJ mol^{-1} K^{-1} in conformational entropy of DHFR accompanies binding of dihydrofolate to form the ternary DHFR·NADPH·dihydrofolate complex. This would correspond to a very unfavorable contribution to the free energy of dihydrofolate binding of roughly +30 kJ/mol^{-1} at 300 K. This result begins to suggest that conformational entropy can have significant impact in the interconversion of kinetic intermediates during enzyme catalysis.

7 Conclusions

The application of NMR relaxation to fundamental issues in enzymology is now reaching into the thermodynamic origins of enzyme catalysis. The development of powerful experimental strategies now allows for the measurement of fast internal dynamics of proteins in various functional states and contexts. The experimental and analytical strategies needed to employ measures of motion as a proxy for the underlying conformational entropy of proteins have begun to mature. Tantalizing new features of both protein motion and the role of conformational entropy in protein function are emerging. Indeed, in the few examples of ligand binding to enzymes studied to date, the role of conformational entropy in defining the energetics of product release appears significant and warrants further detailed examination of other systems. It seems clear that these types of insights are likely not only to impact our general understanding of enzymatic function but also to assist in the more robust design of pharmaceuticals directed against them.

Acknowledgments We gratefully acknowledge the financial support of the National Institutes of Health (DK 39806 and GM 102447) and the Mathers Foundation. We thank Dr. Kathy Valentine and Professor Kim Sharp for continuing discussion.

References

1. Fischer E (1894) Einfluss der confiruation auf die Wirkung der enzyme. Ber Dt Chem Ges 27:2895–2993
2. Koshland DE (1958) Application of a theory of enzyme specificity to protein synthesis. Proc Natl Acad Sci USA 44:98–104
3. Bosshard HR (2001) Molecular recognition by induced fit: how fit is the concept? News Phys Sci 16:171–173
4. Tsai CJ, Ma B, Nussinov R (1999) Folding and binding cascades: shifts in energy landscapes. Proc Natl Acad Sci USA 96:9970–9972
5. Frauenfelder H, Sligar SG, Wolynes PG (1991) The energy landscapes and motions of proteins. Science 254:1598–1603
6. Hilser VJ, Garcia-Moreno EB, Oas TG, Kapp G, Whitten ST (2006) A statistical thermodynamic model of the protein ensemble. Chem Rev 106:1545–1558
7. Henzler-Wildman K, Kern D (2007) Dynamic personalities of proteins. Nature 450:965–972
8. Wolynes PG (2005) Recent successes of the energy landscape theory of protein folding and function. Q Rev Biophys 38:405–410
9. Volkman BF, Lipson D, Wemmer DE, Kern D (2001) Two-state allosteric behavior in a single-domain signaling protein. Science 291:2429–2433
10. Boehr DD, McElheny D, Dyson HJ, Wright PE (2006) The dynamic energy landscape of dihydrofolate reductase catalysis. Science 313:1638–1642
11. Eisenmesser EZ, Bosco DA, Akke M, Kern D (2002) Enzyme dynamics during catalysis. Science 295:1520–1523
12. Eisenmesser EZ, Millet O, Labeikovsky W, Korzhnev DM, Wolf-Watz M, Bosco DA, Skalicky JJ, Kay LE, Kern D (2005) Intrinsic dynamics of an enzyme underlies catalysis. Nature 438:117–121
13. Fraser JS, Clarkson MW, Degnan SC, Erion R, Kern D, Alber T (2009) Hidden alternative structures of proline isomerase essential for catalysis. Nature 462:669, 673
14. Bhabha G, Lee J, Ekiert DC, Gam J, Wilson IA, Dyson HJ, Benkovic SJ, Wright PE (2011) A dynamic knockout reveals that conformational fluctuations influence the chemical step of enzyme catalysis. Science 332:234–238
15. Kamerlin SCL, Warshel A (2010) At the dawn of the 21st century: is dynamics the missing link for understanding enzyme catalysis? Proteins 78:1339–1375

16. Doshi U, McGowan LC, Ladani ST, Hamelberg D (2012) Resolving the complex role of enzyme conformational dynamics in catalytic function. Proc Natl Acad Sci USA 109 (15):5699–5704
17. Dill KA (1990) Dominant forces in protein folding. Biochemistry 29:7133–7154
18. Gilson MK, Given JA, Bush BL, McCammon JA (1997) The statistical-thermodynamic basis for computation of binding affinities: a critical review. Biophys J 72:1047–1069
19. Gilson MK, Zhou HX (2007) Calculation of protein-ligand binding affinities. Annu Rev Biophys Biomol Struct 36:21–42
20. Zhou HX, Gilson MK (2009) Theory of free energy and entropy in noncovalent binding. Chem Rev 109:4092–4107
21. Homans SW (2007) Water, water everywhere – except where it matters? Drug Discov Today 12:534–539
22. Doyle ML (1997) Characterization of binding interactions calorimetry. Curr Opin Biotechnol 8:31–35
23. Tanford C (1978) The hydrophobic effect and the organization of living matter. Science 200:1012–1018
24. Sharp KA, Nicholls A, Fine RF, Honig B (1991) Reconciling the magnitude of the microscopic and macroscopic hydrophobic effects. Science 252:106–109
25. Chandler D (2005) Interfaces and the driving force of hydrophobic assembly. Nature 437:640–647
26. Steinberg IZ, Scheraga HA (1963) Entropy changes accompanying association reactions of proteins. J Biol Chem 238:172–181
27. Cooper A (1976) Thermodynamic fluctuations in protein molecules. Proc Natl Acad Sci USA 73:2740–2741
28. Levitt M, Warshel A (1975) Computer simulation of protein folding. Nature 253:694–698
29. McCammon JA, Gelin BR, Karplus M (1977) Dynamics of folded proteins. Nature 267:585–590
30. Frauenfelder H, Petsko GA, Tsernoglou D (1979) Temperature-dependent X-ray diffraction as a probe of protein structural dynamics. Nature 280:558–563
31. Gurd FR, Rothgeb TM (1979) Motions in proteins. Adv Protein Chem 33:73–165
32. Igumenova TI, Frederick KK, Wand AJ (2006) Characterization of the fast dynamics of protein amino acid side chains using NMR relaxation in solution. Chem Rev 106:1672–1699
33. Karplus M, Ichiye T, Pettitt BM (1987) Configurational entropy of native proteins. Biophys J 52:1083–1085
34. Li Z, Raychaudhuri S, Wand AJ (1996) Insights into the local residual entropy of proteins provided by NMR relaxation. Protein Sci 5:2647–2650
35. Palmer AG, Massi F (2006) Characterization of the dynamics of biomacromolecules using rotating-frame spin relaxation NMR spectroscopy. Chem Rev 106:1700–1719
36. Jarymowycz VA, Stone MJ (2006) Fast time scale dynamics of protein backbones: NMR relaxation methods, applications, and functional consequences. Chem Rev 106:1624–1671
37. Bai YW (2006) Protein folding pathways studied by pulsed-and native-state hydrogen exchange. Chem Rev 106:1757–1768
38. Ottiger M, Bax A (1998) Determination of relative N-H-NN-C′, C-alpha-C′, and C(alpha)-H-alpha effective bond lengths in a protein by NMR in a dilute liquid crystalline phase. J Am Chem Soc 120:12334–12341
39. Henry ER, Szabo A (1985) Influence of vibrational motion on solid-state line-shapes and NMR relaxation. J Chem Phys 82:4753–4761
40. Ferrage F, Pelupessy P, Cowburn D, Bodenhausen G (2006) Protein backbone dynamics through C-13′-C-13(alpha) cross-relaxation in NMR spectroscopy. J Am Chem Soc 128:11072–11078
41. Liu WD, Zheng Y, Cistola DP, Yang DW (2003) Measurement of methyl C-13-H-1 cross-correlation in uniformly C-13-, N-15-, labeled proteins. J Biomol NMR 27:351–364

42. Pelupessy P, Ravindranathan S, Bodenhausen G (2003) Correlated motions of successive amide N-H bonds in proteins. J Biomol NMR 25:265–280
43. Vugmeyster L, Pelupessy P, Vugmeister BE, Abergel D, Bodenhausen G (2004) Cross-correlated relaxation in NMR of macromolecules in the presence of fast and slow internal dynamics. C R Phys 5:377–386
44. Wang TZ, Frederick KK, Igumenova TI, Wand AJ, Zuiderweg ERP (2005) Changes in calmodulin main-chain dynamics upon ligand binding revealed by cross-correlated NMR relaxation measurements. J Am Chem Soc 127:828–829
45. Zhang X, Sui XG, Yang DW (2006) Probing methyl dynamics from C-13 autocorrelated and cross-correlated relaxation. J Am Chem Soc 128:5073–5081
46. Cavanagh J, Fairbrother WJ, Palmer AG, Rance M, Skelton NJ (2006) Protein NMR spectroscopy: principles and practice. Elsevier, Burlington, MA
47. Lipari G, Szabo A (1982) Model-free approach to the interpretation of nuclear magnetic-resonance relaxation in macromolecules. 1. Theory and range of validity. J Am Chem Soc 104:4546–4559
48. Lipari G, Szabo A (1982) Model-free approach to the interpretation of nuclear magnetic-resonance relaxation in macromolecules. 2. Analysis of experimental results. J Am Chem Soc 104:4559–4570
49. Tjandra N, Feller SE, Pastor RW, Bax A (1995) Rotational diffusion anisotropy of human ubiquitin from N-15 NMR relaxation. J Am Chem Soc 117:12562–12566
50. Lee LK, Rance M, Chazin WJ, Palmer AG (1997) Rotational diffusion anisotropy of proteins from simultaneous analysis of N-15 and C-13(alpha) nuclear spin relaxation. J Biomol NMR 9:287–298
51. Pawley NH, Wang C, Koide S, Nicholson LK (2001) An improved method for distinguishing between anisotropic tumbling and chemical exchange in analysis of 15N relaxation parameters. J Biomol NMR 20:149–165
52. Kneller JM, Lu M, Bracken C (2002) An effective method for the discrimination of motional anisotropy and chemical exchange. J Am Chem Soc 124:1852–1853
53. Mandel AM, Akke M, Palmer AG (1995) Backbone dynamics of *Escherichia coli* ribonuclease H1 – correlations with structure and function in an active enzyme. J Mol Biol 246:144–163
54. McIntosh LP, Dahlquist FW (1990) Biosynthetic incorporation of N15 and C13 for assignment and interpretation of nuclear magnetic resonance spectra of proteins. Q Rev Biophys 23:1–38
55. Wand AJ, Bieber RJ, Urbauer JL, McEvoy RP, Gan ZH (1995) Carbon relaxation in randomly fractionally C13 enriched proteins. J Magn Reson B 108:173–175
56. LeMaster DM, Kushlan DM (1996) Dynamical mapping of *E. coli* thioredoxin via C-13 NMR relaxation analysis. J Am Chem Soc 118:9255–9264
57. Lee AL, Urbauer JL, Wand AJ (1997) Improved labeling strategy for C-13 relaxation measurements of methyl groups in proteins. J Biomol NMR 9:437–440
58. Gardner KH, Kay LE (1998) The use of H-2, C-13, N-15 multidimensional NMR to study the structure and dynamics of proteins. Annu Rev Biophys Biomol Struct 27:357–406
59. Goto NK, Gardner KH, Mueller GA, Willis RC, Kay LE (1999) A robust and cost-effective method for the production of Val, Leu, Ile (delta 1) methyl-protonated N-15-, C-13-, H-2-labeled proteins. J Biomol NMR 13:369–374
60. Ollerenshaw JE, Tugarinov V, Skrynnikov NR, Kay LE (2005) Comparison of (CH3)-C-13, (CH2D)-C-13, and (CHD2)-C-13 methyl labeling strategies in proteins. J Biomol NMR 33:25–41
61. Kay LE, Torchia DA, Bax A (1989) Backbone dynamics of proteins as studied by N-15 inverse detected heteronuclear NMR spectroscopy – application to staphylococcal nuclease. Biochemistry 28:8972–8979
62. Farrow NA, Muhandiram R, Singer AU, Pascal SM, Kay CM, Gish G, Shoelson SE, Pawson T, Formankay JD, Kay LE (1994) Backbone dynamics of a free and a

phosphopeptide-complexed Src homology-2 domain studied by N-15 NMR relaxation. Biochemistry 33:5984–6003
63. Muhandiram DR, Yamazaki T, Sykes BD, Kay LE (1995) Measurement of H-2T-1 and T-1p relaxation-times in uniformly C-13-labeled and fractionally H-2-labeled proteins in solution. J Am Chem Soc 117:11536–11544
64. Millet O, Muhandiram DR, Skrynnikov NR, Kay LE (2002) Deuterium spin probes of side-chain dynamics in proteins. 1. Measurement of five relaxation rates per deuteron in C-13-labeled and fractionally H-2-enriched proteins in solution. J Am Chem Soc 124:6439–6448
65. Tugarinov V, Kay LE (2005) Quantitative 13C and 2H NMR relaxation studies of the 723-residue enzyme malate synthase G reveal a dynamic binding interface. Biochemistry 44:15970–15977
66. Liao X, Long D, Li DW, Bruschweiler R, Tugarinov V (2012) Probing side-chain dynamics in proteins by the measurement of nine deuterium relaxation rates per methyl group. J Phys Chem B 116:606–620
67. Ishima R, Petkova AP, Louis JM, Torchia DA (2001) Comparison of methyl rotation axis order parameters derived from model-free analyses of H-2 and C-13 longitudinal and transverse relaxation rates measured in the same protein sample. J Am Chem Soc 123:6164–6171
68. Dellwo MJ, Wand AJ (1989) Model-independent and model-dependent analysis of the global and internal dynamics of cyclosporine-A. J Am Chem Soc 111:4571–4578
69. Tjandra N, Kuboniwa H, Ren H, Bax A (1995) Rotational dynamics of calcium-free calmodulin studied by 15N-NMR relaxation measurements. Eur J Biochem 230:1014–1024
70. Wand AJ, Urbauer JL, McEvoy RP, Bieber RJ (1996) Internal dynamics of human ubiquitin revealed by C-13-relaxation studies of randomly fractionally labeled protein. Biochemistry 35:6116–6125
71. Ishima R, Louis JM, Torchia DA (2001) Optimized labeling of (CHD2)-C-13 methyl isotopomers in perdeuterated proteins: potential advantages for C-13 relaxation studies of methyl dynamics of larger proteins. J Biomol NMR 21:167–171
72. Lee AL, Kinnear SA, Wand AJ (2000) Redistribution and loss of side chain entropy upon formation of a calmodulin-peptide complex. Nat Struct Biol 7:72–77
73. Liu W, Flynn PF, Fuentes EJ, Kranz JK, McCormick M, Wand AJ (2001) Main chain and side chain dynamics of oxidized flavodoxin from cyanobacterium Anabaena. Biochemistry 40:14744–14753
74. Walsh ST, Lee AL, DeGrado WF, Wand AJ (2001) Dynamics of a de novo designed three-helix bundle protein studied by 15N, 13C, and 2H NMR relaxation methods. Biochemistry 40:9560–9569
75. Akke M, Bruschweiler R, Palmer AG (1993) NMR order parameters and free-energy – an analytical approach and its application to cooperative Ca2+ binding by calbindin-D(9 k). J Am Chem Soc 115:9832–9833
76. Yang D, Kay LE (1996) Contributions to conformational entropy arising from bond vector fluctuations measured from NMR-derived order parameters: application to protein folding. J Mol Biol 263:369–382
77. Privalov PL, Gill SJ (1988) Stability of protein structure and hydrophobic interaction. Adv Protein Chem 39:191–234
78. Lee AL, Sharp KA, Kranz JK, Song XJ, Wand AJ (2002) Temperature dependence of the internal dynamics of a calmodulin-peptide complex. Biochemistry 41:13814–13825
79. Law AB, Fuentes EJ, Lee AL (2009) Conservation of side-chain dynamics within a protein family. J Am Chem Soc 131:6322–6323
80. Clarkson MW, Gilmore SA, Edgell MH, Lee AL (2006) Dynamic coupling and allosteric behavior in a nonallosteric protein. Biochemistry 45:7693–7699
81. Clarkson MW, Lee AL (2004) Long-range dynamic effects of point mutations propagate through side chains in the serine protease inhibitor Eglin c. Biochemistry 43:12448–12458

82. Fuentes EJ, Der CJ, Lee AL (2004) Ligand-dependent dynamics and intramolecular signaling in a PDZ domain. J Mol Biol 335:1105–1115
83. Fuentes EJ, Gilmore SA, Mauldin RV, Lee AL (2006) Evaluation of energetic and dynamic coupling networks in a PDZ domain protein. J Mol Biol 364:337–351
84. Popovych N, Tzeng SR, Tonelli M, Ebright RH, Kalodimos CG (2009) Structural basis for cAMP-mediated allosteric control of the catabolite activator protein. Proc Natl Acad Sci USA 106:6927–6932
85. Tzeng SR, Kalodimos CG (2009) Dynamic activation of an allosteric regulatory protein. Nature 462:368–372
86. Frederick KK, Marlow MS, Valentine KG, Wand AJ (2007) Conformational entropy in molecular recognition by proteins. Nature 448:325–329
87. Kahl CR, Means AR (2003) Regulation of cell cycle progression by calcium/calmodulin-dependent pathways. Endocr Rev 24:719–736
88. Yap KL, Kim J, Truong K, Sherman M, Yuan T, Ikura M (2000) Calmodulin target database. J Struct Funct Genomics 1:8–14
89. Marlow MS, Dogan J, Frederick KK, Valentine KG, Wand AJ (2010) The role of conformational entropy in molecular recognition by calmodulin. Nat Chem Biol 6:352–358
90. Marlow MS, Wand AJ (2006) Conformational dynamics of calmodulin in complex with the calmodulin-dependent kinase kinase alpha calmodulin-binding domain. Biochemistry 45:8732–8741
91. Frederick KK, Kranz JK, Wand AJ (2006) Characterization of the backbone and side chain dynamics of the CaM-CaMKIp complex reveals microscopic contributions to protein conformational entropy. Biochemistry 45:9841–9848
92. Blake CCF, Koenig DF, Mair GA, North ACT, Phillips DC, Sarma VR (1965) Structure of hen egg-white lysozyme – a 3-dimensional Fourier synthesis at 2 Å resolution. Nature 206:757–761
93. Koshland DE (1953) Stereochemistry and the mechanism of enzymatic reactions. Biol Rev Camb Philos Soc 28:416–436
94. Vocadlo DJ, Davies GJ, Laine R, Withers SG (2001) Catalysis by hen egg-white lysozyme proceeds via a covalent intermediate. Nature 412:835–838
95. Berger LR, Weiser RS (1957) The beta-glucosaminidase activity of egg-white lysozyme. Biochim Biophys Acta 26:517–521
96. Blake CC, Johnson LN, Mair GA, North AC, Phillips DC, Sarma VR (1967) Crystallographic studies of the activity of hen egg-white lysozyme. Proc R Soc Lond B Biol Sci 167:378–388
97. Rupley JA (1964) The hydrolysis of chitin by concentrated hydrochloric acid, and the preparation of low-molecular-weight substrates for lysozyme. Biochim Biophys Acta 83:245–255
98. Garcia-Hernandez E, Zubillaga RA, Chavelas-Adame EA, Vazquez-Contreras E, Rojo-Dominguez A, Costas M (2003) Structural energetics of protein-carbohydrate interactions: insights derived from the study of lysozyme binding to its natural saccharide inhibitors. Protein Sci 12:135–142
99. Laible NJ, Germaine GR (1985) Bactericidal activity of human lysozyme, muramidase-inactive lysozyme, and cationic polypeptides against *Streptococcus sanguis* and *Streptococcus faecalis* – inhibition by chitin oligosaccharides. Infect Immun 48:720–728
100. Moorman VR, Valentine KG, Wand AJ (2012) The dynamical response of hen egg white lysozyme to the binding of a carbohydrate ligand. Protein Sci 21(7):1066–1073
101. Liu W, Rumbley JN, Englander SW, Wand AJ (2009) Fast structural dynamics in reduced and oxidized cytochrome c. Protein Sci 18:670–674
102. Flynn PF, Bieber Urbauer RJ, Zhang H, Lee AL, Wand AJ (2001) Main chain and side chain dynamics of a heme protein: 15N and 2H NMR relaxation studies of *R. capsulatus* ferrocytochrome c2. Biochemistry 40:6559–6569

103. Maenaka K, Matsushima M, Song H, Sunada F, Watanabe K, Kumagai I (1995) Dissection of protein-carbohydrate interactions in mutant hen egg-white lysozyme complexes and their hydrolytic activity. J Mol Biol 247:281–293
104. Schnell JR, Dyson HJ, Wright PE (2004) Structure, dynamics, and catalytic function of dihydrofolate reductase. Annu Rev Biophys Biomol Struct 33:119–140
105. Epstein DM, Benkovic SJ, Wright PE (1995) Dynamics of the dihydrofolate reductase-folate complex: catalytic sites and regions known to undergo conformational change exhibit diverse dynamical features. Biochemistry 34:11037–11048
106. Schnell JR, Dyson HJ, Wright PE (2004) Effect of cofactor binding and loop conformation on side chain methyl dynamics in dihydrofolate reductase. Biochemistry 43:374–383
107. Mauldin RV, Carroll MJ, Lee AL (2009) Dynamic dysfunction in dihydrofolate reductase results from antifolate drug binding: modulation of dynamics within a structural state. Structure 17:386–394
108. Sawaya MR, Kraut J (1997) Loop and subdomain movements in the mechanism of *Escherichia coli* dihydrofolate reductase: crystallographic evidence. Biochemistry 36:586–603

Top Curr Chem (2013) 337: 95–122
DOI: 10.1007/128_2012_410

Published online: 30 March 2013

Conformational Heterogeneity Within the LID Domain Mediates Substrate Binding to *Escherichia coli* Adenylate Kinase: Function Follows Fluctuations

Travis P. Schrank, James O. Wrabl, and Vincent J. Hilser

Abstract Proteins exist as dynamic ensembles of molecules, implying that protein amino acid sequences evolved to code for both the ground state structure as well as the entire energy landscape of excited states. Accumulating theoretical and experimental evidence suggests that enzymes use such conformational fluctuations to facilitate allosteric processes important for substrate binding and possibly catalysis. This phenomenon can be clearly demonstrated in *Escherichia coli* adenylate kinase, where experimentally observed local unfolding of the LID subdomain, as opposed to a more commonly postulated rigid-body opening motion, is related to substrate binding. Because "entropy promoting" glycine mutations designed to increase specifically the local unfolding of the LID domain also affect substrate binding, changes in the excited energy landscape effectively tune the function of this enzyme *without changing the ground state structure or the catalytic site*. Thus, additional thermodynamic information, above and beyond the single folded structure of an enzyme–substrate complex, is likely required for a full and quantitative understanding of how enzymes work.

T.P. Schrank
Department of Biochemistry and Molecular Biology, University of Texas Medical Branch at Galveston, 301 University Boulevard, Galveston, TX 77555-1068, USA
e-mail: t.parke.schrank@gmail.com

J.O. Wrabl
Department of Biology, The Johns Hopkins University, Mudd Hall Room 117, 3400 N. Charles Street, Baltimore, MD 21218, USA
e-mail: jowrabl@jhu.edu

V.J. Hilser (✉)
Department of Biology, The Johns Hopkins University, Mudd Hall Room 117, 3400 N. Charles Street, Baltimore, MD 21218, USA

T.C. Jenkins Department of Biophysics, The Johns Hopkins University, Mudd Hall Room 117A, 3400 N. Charles Street, Baltimore, MD 21218, USA
e-mail: hilser@jhu.edu

Keywords Adenylate kinase · Conformational fluctuation · Enzyme function · Local unfolding · Protein allostery

Contents

Abbreviations

AK	Adenylate kinase
AK(e)	*E. coli* adenylate kinase
Ala	Alanine
AMP	Adenosine monophosphate
AMPbd	AMP binding domain of adenylate kinase
Ap5A	P^1,P^5-di(adenosine-5′) pentaphosphate
ATP	Adenosine triphosphate

BC	Binding competent
BI	Binding incompetent
CD	Circular dichroism spectroscopy
CORE	Core domain of adenylate kinase
CPMG	Carr-Purcell-Meiboom-Gill
Gly	Glycine
HSQC	Heteronuclear single quantum coherence
Ile	Isoleucine
ITC	Isothermal titration calorimetry
K	Equilibrium constant
LID	LID domain of adenylate kinase
NMR	Nuclear magnetic resonance spectroscopy
U	Unfolded
Val	Valine
ΔG	Gibbs free energy
ΔS	Entropy change

1 Introduction: The Invisible Excited State Energy Landscape Is Generally Important for Enzyme Allostery and Function: Is This Knowledge Relevant to Understanding *Escherichia coli* Adenylate Kinase?

1.1 An Emerging Relationship Between Protein Conformational Fluctuations and Enzymatic Function

Enzymes are ubiquitous catalysts for the chemical reactions of life [1–3]. In disease states resulting from altered chemistry, enzymes are often the targets of bioactive compounds and commercially available pharmaceuticals. As such, the fundamental importance of detailed and predictive understanding of enzymatic function is generally critical to the effectiveness of medicine as well as increased knowledge of the underlying biology. Much progress has been made in understanding the structure and mechanism of some enzymes, currently culminating in the ability to alter rationally function in limited favorable cases [2–4]. One outstandingly unique example of the successful application of understanding is the fully computational design of a novel enzyme [5].

Such progress has been incrementally achieved, building on more than a century of studies focused on enzyme reaction kinetics and the thermodynamic properties of enzyme–substrate interactions [6]. However, a predictive chemical and physical understanding of such parameters is limited, with insights being gained mostly from biological investigations of enzymatic adaptation [7, 8]. The present inability to modulate consistently and rationally classical enzymological observables by mutation highlights the true dearth of our collective understanding of such basic phenomena.

An emerging consensus in the field is that the lack of understanding of native state conformational fluctuations (i.e., dynamics) is at least one of the critical barriers to the physical understanding of enzymatic function [9]. Of late, numerous investigations have taken up this challenge [10–14]. Accepting this consensus, however, much remains unknown concerning the nature of the ensemble of interconverting functionally relevant conformations and the transition states governing the kinetics of these interconversions [15, 16].

1.2 Insights into Proteins from Investigating Conformational Fluctuations in the Native State Ensemble

One approach to addressing this issue (and the one adopted by our laboratory) has been to consider the functionally relevant native state of an enzyme to be approximated by the set of all possible equilibrium microstates generated by local unfolding of the protein [17–19]. The approach has proved to be successful in recapitulating, and in several cases quantitatively predicting diverse and seemingly unrelated properties of protein native states, including hydrogen exchange protection factors [20], allosteric communication between domains [21, 22], and global stability effects of pH [23] and denaturant [24].

Importantly, these investigations have given credence to the non-intuitive possibility that an ensemble of mostly folded (but containing segments that are highly disordered) states, experimentally indistinguishable from the equilibrium local unfolding model used in our theoretical approach, may indeed be critical for enzyme function and adaptation. The importance of these observations is twofold. First, important processes such as allostery can occur in the absence of a pathway of intraprotein communication that can be gleaned from analysis of the high resolution structure [21, 22, 25, 26], a strategy that dominates the biophysical analysis of function. Second, and equally important, changes in function can be modulated by changing the stability in different parts of the molecule, independent of the nature of the change, a phenomenon that is exemplified by the ability of surface mutations, for example, to modulate activity [27, 28]. As a consequence, the susceptibility of functional changes will be heterogeneously distributed throughout the protein in a way that may stand counter to intuitive notions of work being propagated in a mechanical way through the protein [29].

With regard to fluctuations and enzyme function, a still rich set of studies by Somero and colleagues [8, 30] has demonstrated the importance of enzyme conformational fluctuations through purely evolutionary considerations. In general, these studies have illuminated the amazing ability of enzymatic adaptation to recapitulate precisely the homeostatic K_m and k_{cat} at a wide variety of physiological temperatures. Intuitively, this process of adaptation may be imagined to involve point mutations that preserve similar active site structure in order to preserve the chemical function. Somero's seminal contribution was the discovery of glycine (Gly) point mutations in lactate dehydrogenase that served to adapt the enzyme to psychrophilic conditions [8]. Indeed, these mutations were located outside the active site, suggesting conserved ground state structure and chemistry. Such mutants were hypothesized to function by adding more conformational entropy to the native state ensemble, implying that conformational states other than the ground state structure conferred the temperature adaptation. Given these observations, the natural questions then become: which conformational states besides the ground state are functionally relevant, and how can these invisible excited states be studied without affecting the catalytic activity of the ground state?

Inspired by such questions, our group has been working towards a comprehensive understanding of the native state of enzymes, and proteins in general, as the Boltzmann statistically weighted set of all possible conformational states of local unfolding. In the extreme cases, local unfolding of all residues within all states is equivalent to global unfolding of the molecule, and no local unfolding within any state is equivalent to a single ground state conformation, such as the X-ray crystal structure. In our view, quantitation of the populations of conformational states that exist between these two extremes, which have been largely ignored in most other work, turn out to be crucial for understanding how enzymes may function, adapt, and evolve [31]. In the following chapter we elaborate on experimental results [27] that document local unfolding in the native state ensemble of adenylate kinase (AK) and the implications thereof for biological function (substrate binding and possibly catalysis) and evolutionary adaptation [28].

1.3 Domain Structure of Adenylate Kinase: How Relevant Is the LID Domain to Structure and Function?

AK has been studied in detail for decades by many groups, and much is known about its structure, function, chemical kinetics, and taxonomy [11, 32–38]. AK catalyzes the reversible Mg^{2+} dependent phosphoryl transfer reaction ATP + AMP $\leftrightarrow$ 2ADP [39], and consequently plays a primary role in maintaining proper cellular energy balance.

The enzyme has a structure composed of up to three domains: a "CORE" domain important for overall stability, which also contains the catalytic residues critical to phosphoryl-transfer, an "AMPbd" domain containing the AMP and ATP substrate binding sites, and an optional "LID" domain that increases catalytic efficiency by covering the binding sites (Fig. 1a). Although the known structures document conformational heterogeneity of the LID domain (Fig. 1b), central to most experimental interpretations [11, 12, 32] and computational simulations [40–46] are

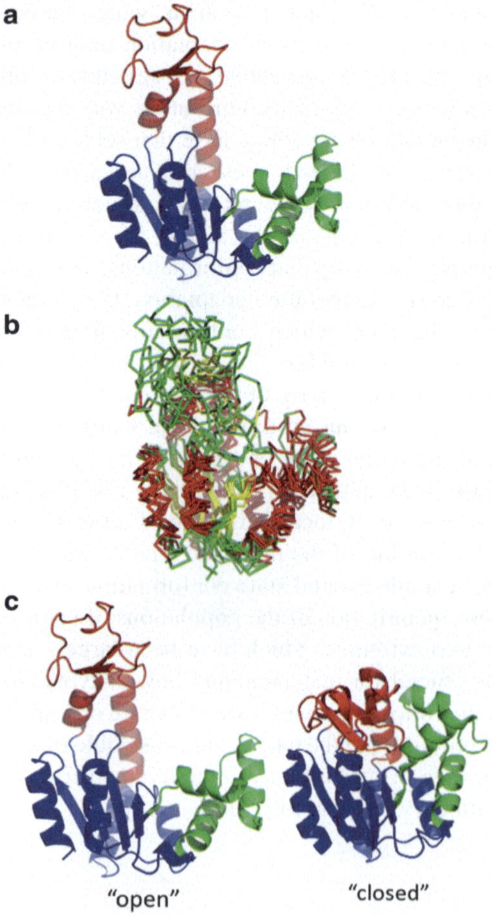

Fig. 1 Domain structure and conformational heterogeneity in *E. coli* adenylate kinase [AK(e)]. (**a**) The enzyme is composed of three domains: a CORE domain (*blue*, residues 1–28, 73–112, 177–214, PDB ID 4AKE), a AMPbd domain (*green*, residues 28–72), and a LID domain (*red*, residues 113–176). (**b**) Multiple superposition of various known *apo* structures of AK demonstrating conformational heterogeneity of the LID. PDB structures 1AK2, 1E4V, 2AR7, 2RH5, 2XB4, 3GMT, 3NDP, each pairwise superimposed on 4AKE, the AK(e) structure are displayed. Helices are colored *red*, strands *yellow*, and loops *green*. Optimal pairwise superpositions of each structure with *apo* AK(e) (PDB ID 4AKE) were performed with TM-align [70] and displayed with pyMOL (Schrödinger, Inc., New York). (**c**) Commonly assumed "open" and "closed" structures of AK(e), based on PDB structures 4AKE and 1AKE, respectively. Coloring is identical to (**a**)

X-ray crystal structures of an "open" state [33] where the LID domain is away from the AMPbd domain and a "closed" state [34] where the LID domain closely contacts the AMPbd and CORE domains, enclosing substrate (Fig. 1c). In particular, most computational work treats the reaction coordinate of AK as a hinged rigid-body movement between the open and closed states. As will become clear in the

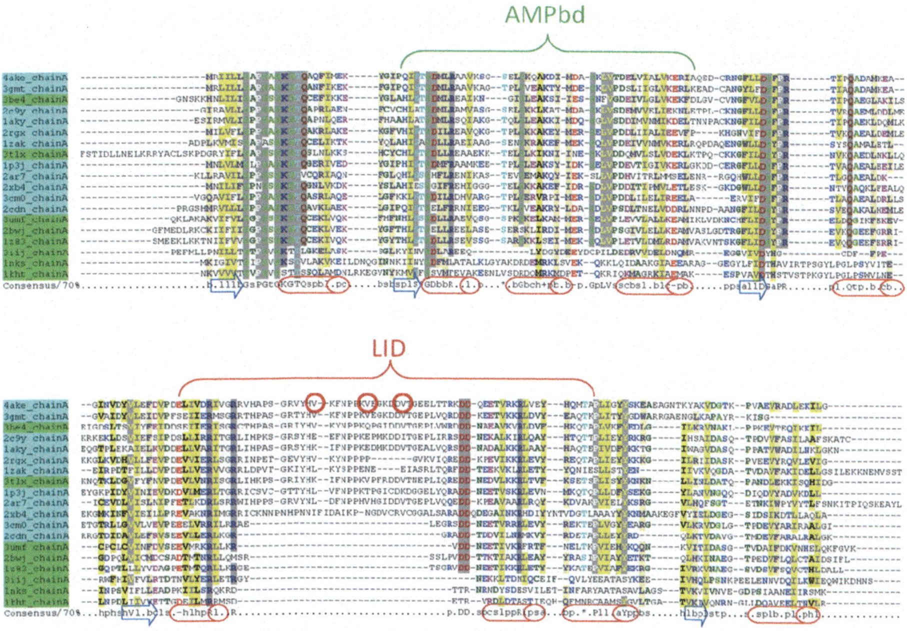

Fig. 2 Well known diversity of homologous AK amino acid sequences: "short" and "long" forms. All adenylate kinase structures were retrieved from the Protein Data Bank (PDB) [71]. Amino acid sequences corresponding to each structure were extracted and clustered at 60% identity using the CD-HIT webserver [72]. Representatives from each cluster were aligned using PROMALS3D [73] and colored with CHROMA [74]. Locations of structural helices are marked with *red cylinders* and strands by *blue arrows*. Prokaryotic or eukaryotic mitochondrial and chloroplast sequences are indicated by *cyan* and eukaryotic cytosolic sequences by *green*. Domain structure referred to in Fig. 1a is labeled. Sites of three "entropy-promoting" valine to glycine mutations in the LID domain of AK(e), discussed in detail later in the text, are indicated by *red circles*

following analysis, our experimental data [27, 28] directly challenge this popular interpretation for the *Escherichia coli* form of the enzyme, AK(e).

While not frequently addressed, taxonomic evidence also calls into question the importance of rigid-body LID movement for the reaction proper. Comparison of amino acid sequences and crystal structures for AK from many species demonstrates that the LID domain is not essential for a functional enzyme [38, 47]. Although there are exceptions, notably for mitochondrial AK [37], an organizing trend in these data is that eukaryotic AKs lack some or the entire LID, while prokaryotic forms contain the LID (Fig. 2). Thus, an interesting speculation is that an optional LID domain is an evolutionary innovation, of marginal importance for catalytic function per se but of great importance for regulation or "fine-tuning" of the existing function. In other words, the rigid opening and closing of a catalytically non-essential LID, the focus of so much computational attention, may not necessarily be investigating the most biologically important adaptation. This proposal is firmly supported by the experimental evidence for functionally relevant local unfolding in AK(e), described below.

2 Revealing an Enzyme's Invisible Excited State Energy Landscape: Deploying Surface Glycine Mutations as Selective Probes Decoupling the Ground State from Temperature-Dependent Fluctuations

2.1 A Creative Mutation Strategy Based on Entropic Stabilization of Excited Conformational States

In order to investigate how similar native state conformational fluctuations are to unfolding reactions, and to uncover the functional significance of such local unfolding for binding or catalysis, consider the following set of experiments designed to amplify and examine fluctuations in AK(e). The goal of this mutation strategy was to probe selectively the conformational manifold of states that comprise various states of the LID domain, in search of states that both resemble local unfolding and are relevant to substrate binding. To accomplish this goal, point mutations were carefully chosen to promote excited states without disturbing the structural or catalytic properties of the fully folded ground state. As discussed above, this mutation strategy could be considered an in vitro reflection of a natural strategy observed in the functional adaptation of lactate dehydrogenase [8].

In principle, the energetic promotion of unfolded or excited states could be affected purely by changes in the denatured state itself, with no energetic or structural perturbation to the folded conformation. A good example of this principle is the thermodynamic mechanism by which denaturants destabilize protein folds [48]. Urea, for example, stabilizes both the native and denatured states of proteins in water. The apparent destabilization of the folded conformation results from the greater stabilization of the denatured state, due to the much larger amount of surface area exposed therein [48]. To effect such a change by point mutation, the specific properties of amino acids that are greatly expressed in the denatured state and weakly expressed in the folded state must be considered. In the denatured state, peptides quasi-randomly search through available conformational degrees of freedom because of the presence of many rotatable bonds. The restriction of this conformational freedom is manifest as the large unfavorable entropy of protein folding [49]. The beta-branched amino acids valine (Val) and isoleucine (Ile) are known to have a relatively small decrease in conformational entropy upon unfolding, due to steric collisions of their bulky side chains in the denatured state [49, 50]. The absence of a side chain in Gly, on the other hand, results in a comparatively large *increase* in the conformational entropy of unfolding [49]. Therefore, Val to Gly point mutations should, in principle, decrease the stability of a protein's folded structure by increasing the conformational entropy of the denatured state.

Unlike structure-based mutagenesis that stabilizes or destabilizes multiple interactions within the folded state of a protein, one advantage of this selective denatured state approach is that the expected stability change can be quantitatively

Table 1 Simulation of the relative conformational space available upon glycine mutation in a disordered peptide

Host–guest pentapeptide pair	$\Omega_{X\text{-}Gly}$	$\Delta\Delta S_{X\text{-}Gly}$ (cal/(mol K))
[AAAAA]–[AAGAA]	3.03	2.20
[AAVAA]–[AAGAA]	4.28	2.89

All simulations were performed using MPMOD software [28, 51, 52]. Equal probabilities were assumed for all valine rotamers. Atomic radii were taken from published values [76–78]

estimated. Moreover, as the change is entirely due to conformational entropy, the stability change can actually be accurately simulated using a conceptually simple algorithm based on hard-sphere collision of atomic radii and well known geometric features of polypeptides such as bond lengths and angles [28, 51, 52]. Such a simulation was performed for pairs of unstructured host pentamers in which the central residue was either Gly or not-Gly (e.g., Ala or Val). An ensemble of denatured state conformations was constructed for each pentamer by sampling a uniform distribution of φ and ϕ torsions n_{test} times and retaining only those n_{allowed} conformations that did not sterically clash [28]. Then, the ratio of available conformational space for each pentamer was calculated as described previously [53]:

$$\Omega_{\text{Ala-Gly}} = \frac{\left[n_{\text{allowed,Ala}}/n_{\text{test,Ala}}\right]}{\left[n_{\text{allowed,Gly}}/n_{\text{test,Gly}}\right]} \tag{1}$$

The expected change in entropy could then be estimated through the expression [53]

$$\Delta\Delta S_{\text{Ala-Gly}} = -R\ \ln\left[\Omega_{\text{Ala-Gly}}\right] \tag{2}$$

According to this analysis, the conformational entropy difference obtained when Ala is mutated to Gly is 2.2 cal/(mol K) (Table 1), a value similar to that measured from calorimetric protein unfolding experiments (2.3–2.7 cal/(mol K) [49]). Simulations involving Val result in an even larger gain of entropy (Table 1). For this latter mutation, the difference is approximately 0.9 kcal/mol at 25°C. In short, these studies reveal that mutations either from Ala or Val to Gly can affect the conformational entropy of a disordered state by almost 1.0 kcal/mol, even if the change does not affect the stabilizing non-covalent interactions in either state. This observation can be put to use in the search for locally unfolded states in AK(e).

Having a substantial expected entropy change, Val to Gly mutations at various surface exposed positions in the AK(e) LID domain were selected for experimental test. As discussed, this type of mutation would be expected to increase the conformational entropy associated with unfolding this region of the protein, if indeed locally unfolded states were already significantly populated in the native state ensemble. Several objective criteria were defined for selecting mutation positions, such that the expected energetic effects would exclusively perturb the denatured state conformational entropy rather than the folded protein structure. These criteria

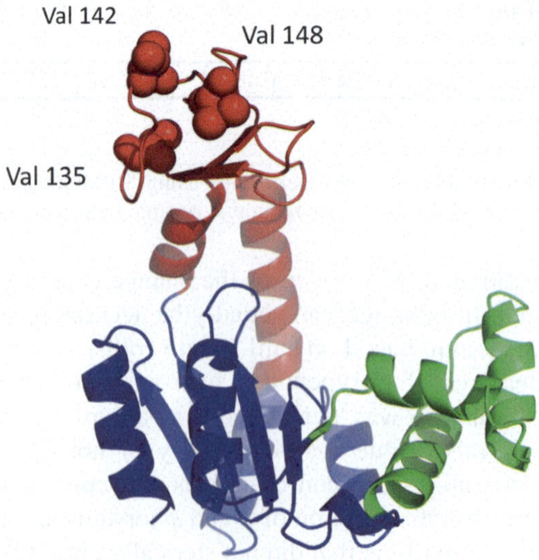

Fig. 3 Three surface exposed valine to glycine entropy promoting mutations in the AK(e) LID domain. Domain colors are as given in Fig. 1a. The locations of the mutations, not expected to affect the folded ground state structure of the enzyme, are shown as *space filling spheres*

were (1) the side chains of the mutated residues were highly surface exposed, involving no or few intermolecular contacts, (2) the side chains did not contain charged or polar groups that might contribute to important hydrogen bonds or long-range electrostatics, and (3) the mutated residues were at least 8 Å distant from the active site, thus ensuring no contact with ligand. Ideally, such a strategy would be expected not to disturb any stabilizing contacts in the folded structure, not to disturb any contacts between enzyme and ligand, and not to disturb essential catalytic chemistry. Thus, it was also expected that little or no perturbation would occur to the structure of the bound complex. Although other positions in the AK(e) LID domain could potentially satisfy these requirements, three positions were selected for further study: Val 135, Val 142, and Val 148 (Fig. 3). These three surface exposed valines were experimentally mutated to Gly.

2.2 Entropy Promoting Mutations Do Not Perturb the Enzyme Ground State Structure of AK(e)

Did the entropy-promoting mutations actually preserve the ground state structure of AK(e)? Careful comparison of X-ray structures of the wild-type enzyme bound complex with the mutant V148G bound complex, demonstrates an affirmative answer to this question. Such a detailed comparison was possible due to the fortunate crystallization of both systems in the same ($P2_12_12$) space group and asymmetric unit. The all-atom RMSD between wild-type and mutant structures is quite small, approximately 0.2 Å. This value was, surprisingly, less than the approximately 0.4 Å RMSD obtained between the two distinct molecular copies

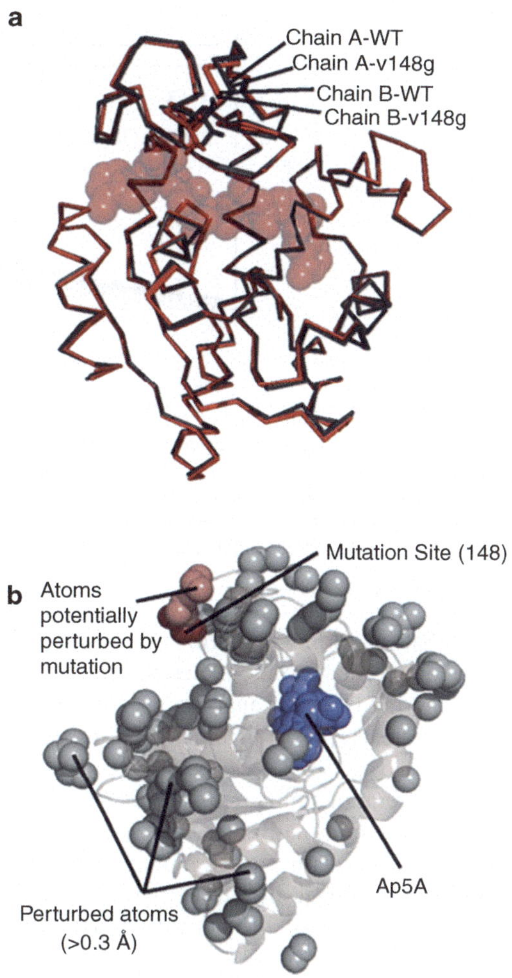

Fig. 4 The ground state structure of AK(e) is unaffected by entropy-promoting LID mutations. (**a**) Structural alignment of the experimentally determined crystal structures of wild-type (PDB ID 3HPQ) and mutant V148G (PDB ID 3HPR). Shown in *red* are chains of each protein from positions A of their respective asymmetric units, and shown in *black* are chains from positions B. Substrate analog Ap5A is shown in *spacefill*. (**b**) Analysis of possible structural perturbations due to mutation. The *gray spheres* represent all atoms that move >0.3 Å from the wild-type to the mutant structure in both copies within the asymmetric unit. The *dark red spheres* show the mutation site V148G. The *light red spheres* show all perturbed atoms (*gray*) that can be connected to the mutation site by a continuous chain of no more than 6 Å per step of other perturbed atoms. Ap5A is shown in *blue spacefill*

in the respective asymmetric units (Fig. 4a). Crystal packing effects on each structure could be quantitatively assessed and were also found to be minimal (Fig. 4a). Excluding such artifacts, atomic positions that were potentially different (>0.3 Å) between the optimally aligned structures of V148G and wild-type are highlighted in Fig. 4b. According to the figure, the mutation site clearly did not exhibit increased structural differences relative to the rest of the molecule. In particular, the active site region around the Ap5A was not disproportionally affected by structural changes. Also, most of the atoms that did show small structure differences were localized to surface exposed side chains (Fig. 4b), where increased mobility likely resulted in a poorer determination of electron density. Within the experimental resolution, it appeared that the mutation strategy was indeed successful at preserving the ground state structure.

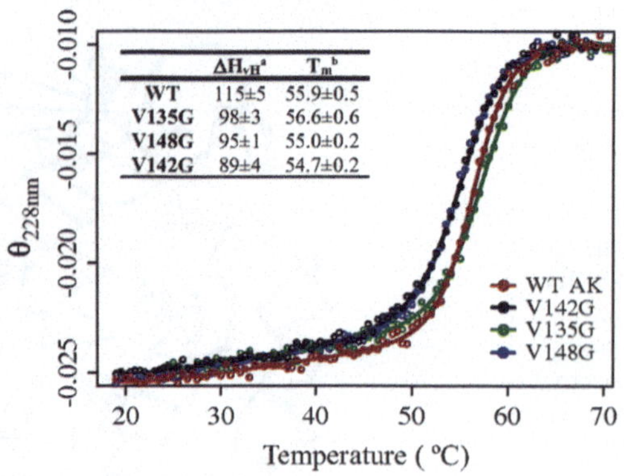

Fig. 5 Thermal unfolding monitored by circular dichroism for wild-type AK(e) and mutants, global stability was minimally affected. Signal was monitored at 228 nm and is given in units of (degrees cm^2)/(dmol res). Denatured state signals are normalized to a common value to aid interpretation. Smooth lines result from fits of the data points to a two-state thermodynamic model. The *inset table* shows thermodynamic parameters obtained from the two-state fits: superscript *a* is the van't Hoff enthalpy in kcal/mol at the transition midpoint temperature T_m (superscript *b*)

Global unfolding experiments monitored by circular dichroism (CD) also corroborated that the ground state structure was preserved for these mutants. Two state thermodynamic analysis of the data revealed that the apparent melting temperatures (T_m) of all mutants were decreased by less than 1.5°C relative to wild-type (Fig. 5). These results are consistent with the expected effects of the entropy promoting mutations in two ways. First, the global stability, as estimated by the T_m, was minimally affected by mutations (Fig. 5, inset), suggesting that the ground state structures of all mutants were minimally affected. Second, all mutants slightly destabilized the structure relative to wild-type, as predicted if the conformational entropy of the denatured states were actually increased, as described above.

3 The AK(e) Energy Landscape: Experimental Evidence for Local Unfolding of the LID Domain

3.1 Exchange Broadening Observed in 2D NMR Spectrum of the Mutant LID Domain

If indeed these mutations minimally perturbed the ground state of AK(e), how could their effects on excited states be described and quantified? One clue came from comparison of the NMR ^1H–^{15}N HSQC spectra of wild-type and the mutant V142G, both in the absence of ligand. These data provided a site-resolved view of

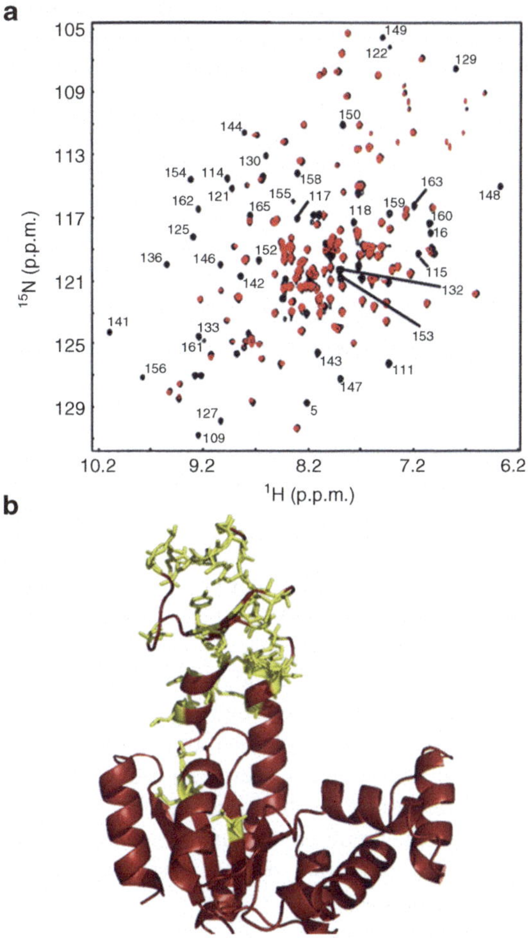

Fig. 6 NMR results implicating LID domain effects due to entropy promoting mutations. (**a**) ^{1}H–^{15}N HSQC spectra of wild-type (*black*) and mutant V142G AK(e) at 33°C. *Labels* indicate those residues that were assignable from published values [75] and also exhibited measurable intensity in the wild-type spectrum but no measurable intensity in V142G. (**b**) The same residues labeled in Fig. 6a are shown as *yellow sticks*, projected onto the "open" conformation of AK(e) (PDB ID 4AKE)

possible structural or dynamic changes caused by the mutation. In agreement with the crystallographic results, most amide resonances throughout the wild-type and mutant proteins exhibited respective peaks with similar location and intensity (Fig. 6a), suggesting overall similarity of the ground state conformation. However, closer inspection of these data indicated that a number of peaks essentially disappeared in the mutant spectrum (Fig. 6a). Interestingly, no new resonances appeared in the mutant, and the missing resonances reappeared at lower temperature (data not shown), suggesting both that the ground state structure was not altered and that no new structure, such as a "closed" LID domain, was substantially populated as a result of the mutation. The most probable explanation for these observations was that a chemical exchange process had broadened the peaks beyond detection. All except two of the broadened and assignable residues were consecutive in sequence from position 109 to 165 (Fig. 6b). This suggested that the mutation was affecting a large region of the protein, encompassing the LID domain. We also observed that the HSQC spectra of the other two mutants, V135G and V148G, were qualitatively similar in terms of peak broadening in the LID domain (data not shown). This last result might also be expected if all three mutants were involved in the same physical process, in other words all promoting the same gain of denatured state conformational entropy in the LID domain.

3.2 NMR Relaxation Dispersion Is Not Consistent with Rigid-Body Opening and Closing of the LID Domain

To explore the possibility that the entropy promoting mutants affect a conformational exchange process, NMR relaxation dispersion experiments (^{15}N Carr-Purcell-Meiboom-Gill, CPMG [54–56]) were performed on the V142G protein. Under the assumption of a two state conformational exchange process on the microsecond to millisecond time scale, CPMG profiles can estimate the chemical shifts of the two conformations, the populations of each conformation, and the rates of interconversion between them. (It is noted in passing that these experiments have been used with good effect to measure other properties of AK(e), such as the conformational change under saturation of ligand, and the conformational transition relevant to product release [11].)

In general, conformational transitions between two kinetically distinct substates A and B in a protein's energy landscape result from changes in atomic motions. If these transitions are both fast relative to the NMR timescale and occur between different magnetic environments such that a chemical shift difference $\Delta\omega$ can be detected, populations of the states p_A and p_B can be estimated from the following basic equations [57]:

$$R_{2,\mathrm{eff}} = R_{20} + R_{\mathrm{ex}} \tag{3}$$

$$R_{\mathrm{ex}} \sim \frac{p_A\, p_B\, \Delta\omega^2}{k_{\mathrm{ex}}} \tag{4}$$

In these equations, R_{20} is the intrinsic relaxation in the absence of chemical exchange, R_{ex} is the chemical exchange contribution to $R_{2,\mathrm{eff}}$, the observed transverse relaxation rate, and k_{ex} is the sum of the forward (k_{AB}) and reverse (k_{BA}) rate constants.

Variables p_A, p_B, k_{AB}, and k_{BA} were extracted from relaxation–dispersion experiments using the procedure described by Vallurupalli and Kay [58] as summarized in the following. First, $R_{2,\mathrm{eff}}$ was measured for each resonance from the difference in intensities between spectra collected in the absence ($I_{\mathrm{reference}}$) and presence ($I_{\nu\mathrm{CPMG}}$) of a constant time relaxation element of duration T:

Table 2 CPMG results for LID domain residues of apo V142G AK(e)

Residue	$k_{\text{unfolding}}$ (s^{-1})[a]	k_{folding} (s^{-1})[a]	P_{unfolded} (%)[b]	$\Delta\omega$ (ppm)[c]	$\langle\chi^2\rangle/N$[d]
125	18 ± 1	570 ± 100	3.0 ± 0.6	3.29	1.4 ± 0.4
130	31 ± 3	670 ± 46	4.4 ± 0.4	1.70	1.9 ± 0.0
141	28 ± 5	600 ± 120	4.5 ± 0.1	2.35	1.2 ± 0.3
142	13 ± 1	140 ± 36	8.5 ± 0.2	2.08	2.8 ± 0.3
143	24 ± 1	660 ± 22	3.5 ± 0.1	2.86	3.0 ± 1.7
146	12 ± 4	200 ± 150	5.7 ± 4.5	1.30	1.1 ± 0.7
148	20 ± 1	290 ± 72	6.5 ± 0.1	4.60	1.4 ± 0.4
159	19 ± 6	530 ± 430	3.5 ± 0.3	4.10	0.7 ± 0.1
Global fit	19 ± 0	420 ± 27	4.3 ± 0.3	N/A	2.0 ± 0.4

[a]Average rate constants and standard error from three direct repeats of complete experimental and fitting procedure (TP Schrank, A Majumbar, unpublished results). Experimental conditions were 19°C in the absence of substrate

[b]Calculated from the presumed rate constants, with propagated standard error

[c]The expected change in chemical shift upon unfolding this residue, based on random coil values [60, 79]

[d]Average and standard deviation of the reduced chi-squared value for the three individual fits

$$R_{2,\text{eff}} = \frac{1}{T} \ln\left(\frac{I_{\text{vCPMG}}}{I_{\text{reference}}}\right) \tag{5}$$

Next, the individual contributions to $R_{2,\text{eff}}$, as expressed in the Bloch–McConnell equations [59] were obtained by minimizing the chi-squared value (χ^2) of the following equation:

$$\chi^2 = \sum_{\text{vCPMG}} \frac{\left[R_{2,\text{eff}}(\text{vCPMG}) - R_{2,\text{eff}}^{\text{calc}}(\text{vCPMG})\right]^2}{\left[\sigma_{R_{2,\text{eff}}}(\text{vCPMG})\right]^2} \tag{6}$$

In (6), $R_{2,\text{eff}}$ is the measured value from (5) and $\sigma_{R2,\text{eff}}$ is the error in said value, estimated by propagating through (5) the standard deviation of the average noise observed in the spectra. $R_{2,\text{eff}}^{\text{calc}}$ was calculated by numerical solution of the Bloch–McConnell equations, assuming ideal CPMG refocusing pulses, with the summation over all increments of vCPMG. Finally, values of k_{AB} and k_{BA} were calculated from the values of k_{ex} and p_{B} obtained from the fitting results of (6) according to the relations $k_{\text{AB}} = k_{\text{ex}}/p_{\text{B}}$, $k_{\text{ex}} = k_{\text{AB}} + k_{\text{BA}}$, and $k_{\text{BA}} = k_{\text{ex}}/p_{\text{A}}$. In this analysis, states A and B corresponded to conformational states of the LID domain such that state A was folded and state B was unfolded, as noted in Table 2.

Under *apo* conditions, the CPMG data for V142G revealed two interesting findings. First, all of the residues in the LID domain could be well fitted (either individually or globally) to a common two state process (Fig. 7, Table 2). Importantly, the $\Delta\omega$ used in these fits assumed a transition between the chemical shifts for the folded AK(e) and random coil chemical shifts for all residues [60]. This observed common process was therefore consistent with an equilibrium between the wild-type folded structure and a random-coil-like unfolded structure. Second,

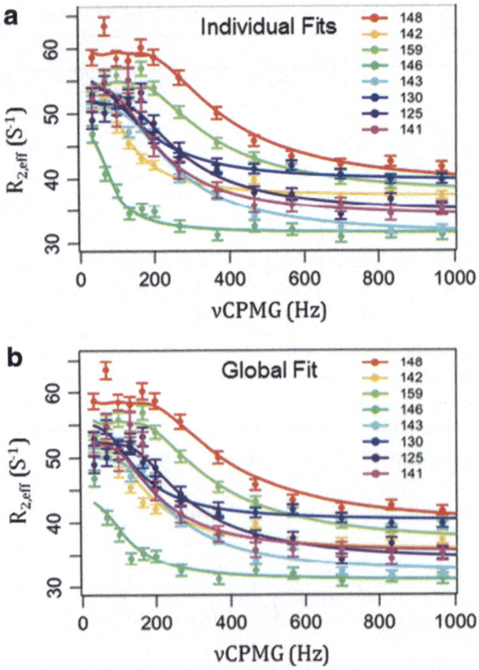

Fig. 7 CPMG results for *apo* V142G are consistent with all-or-none local unfolding of the LID domain. (**a**) Individual fits of the data, i.e., data points for each residue are fitted to (6) to obtain many sets of fitted parameters. These fitted parameters are individually listed in Table 2. (**b**) Global fit of all data points from all residues to a hypothetical two state transition between folded AK(e) and random coil chemical shifts. This one set of fitted parameters is given in Table 2 in the row labeled "Global". Relaxation dispersion data was collected for *apo* V142G AK(e) at 19°C in 50 mM MOPS, 50 mM NaCl, 20 mM $MgCl_2$, 5% glycerol, 8 mM DTT, 8 mM TCEP, pH 7.85. Data were collected at both 800 and 600 MHz field strength; 800 MHz results are displayed

the unfolded LID domain was significantly populated to a level of approximately 3–9% in the V142G mutant at 19°C (Table 2). These results are particularly striking because they suggested that the LID domain could independently unfold to a random-coil-like state, in stark contrast to the usual depiction of AK(e) LID motion as a rigid-body open and closing.

3.3 *Local Unfolding of the AK(e) LID Domain Is Essential for Substrate Binding*

The "closed" conformation of AK(e) (Fig. 8) suggests that access of substrate to the binding site is sterically blocked relative to the "open" conformation. Could a local unfolding event like that observed (and apparently affected by the surface exposed mutations) be important for substrate binding and possibly catalytic function?

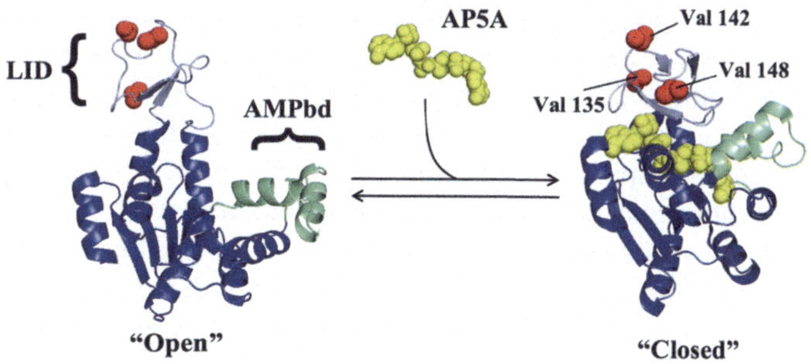

Fig. 8 Schematic equilibrium of binding Ap5A to structurally open and closed AK(e). Global conformations, domain structure, and locations of entropy-promoting LID domain mutations are displayed. Bisubstrate analog P[1],P[5]-di(adenosine-5′) pentaphosphate (Ap5A) is shown in *yellow*. Note that even in the global "closed" conformation the locations of all three valines are solvent exposed

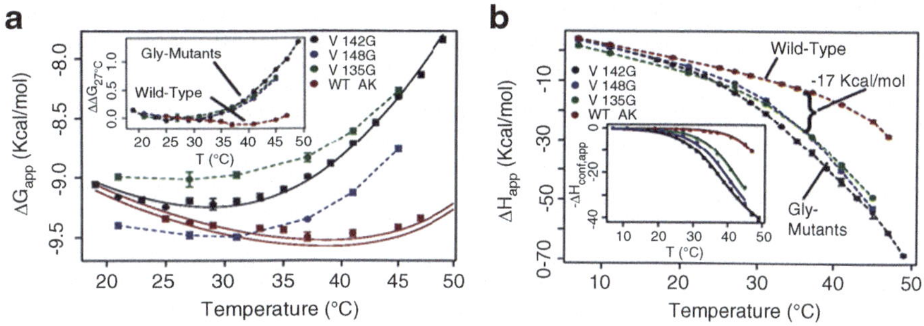

Fig. 9 Surface mutations affect the thermodynamics of binding. (**a**) Apparent free energy (ΔG_{app}) of binding substrate analog Ap5A. *Solid black line* shows the fitted curve for V142G data. *Solid red lines* represent the prediction of wild-type data based on the fitting of $\Delta H_{conf,app}$. *Error bars* show the average standard error of the prediction. The *inset* displays the change in ΔG_{app} for each protein referenced with respect to 27°C: $\Delta\Delta G_{27°C}(T) = \Delta G(T) - \Delta G(27°C)$, clearly demonstrating the similar effects of each mutant on binding. (**b**) Apparent enthalpy of binding (ΔH_{app}) Ap5A; similar changes are evident for all mutants. The *inset* shows corrected data and fits for $\Delta H_{conf,app}$ (6)

To ascertain the impact of the mutations on substrate binding, isothermal titration calorimetry (ITC) experiments were used to measure the binding between AK(e) and the bisubstrate analog P[1],P[5]-di(adenosine-5′) pentaphosphate (Ap5A, Fig. 8).

All three mutants resulted in substantial changes in the thermodynamics of binding. At 37°C and above, all mutations decreased binding affinity of the substrate analog (Fig. 9a), while simultaneously effecting large increases in the favorable enthalpy of binding (Fig. 9b). The observed temperature dependencies of the binding affinities were similar among all the mutants but different from wild-type AK(e) (Fig. 9a, inset). The binding enthalpies were also similar among the mutants but different from wild-type (Fig. 9b). Taken together with the global stability and local unfolding results described in previous sections, all three mutants were likely manifesting these functional changes by a similar mechanism.

The most straightforward interpretation of the ITC results is to postulate that the binding reaction is coupled to a conformational equilibrium that could be modulated by the entropy-promoting mutations. An expression articulating this idea is an association reaction where a "binding incompetent" (BI) state, having little or no affinity for substrate, is in equilibrium with a "binding competent" (BC) high-affinity state. The addition of substrate promotes the population of the BC state through mass action [61]. The apparent free energy would thus consist of two terms [27]:

$$\begin{aligned}\Delta G_{\mathrm{app}}(T) &= \Delta G_0(T) - \Delta G_{\mathrm{conf,app}}(T)\\ &= -RT\ \ln K_0(T) + RT\ \ln(1 + K_{\mathrm{conf}}(T))\end{aligned} \tag{7}$$

In (7) the first term represents the intrinsic free energy of interaction between the BC state and the substrate and the second term accounts for the apparent contribution of the conformational equilibrium between the BI and BC states to the free energy.

Similarly, the binding enthalpy of the equilibrium also has two terms [27]:

$$\Delta H_{\mathrm{app}}(T) = \Delta H_0(T) - \Delta H_{\mathrm{conf,app}}(T) = \Delta H_0(T) - \frac{K_{\mathrm{conf}}(T)}{1 + K_{\mathrm{conf}}(T)}\Delta H_{\mathrm{conf}}(T) \tag{8}$$

The conformational term in (8) can be used to fit directly the ITC data (Fig. 9b, inset) for the thermodynamic parameters governing the conformational equilibrium ($T_{\mathrm{m,conf}}$, $\Delta C_{\mathrm{p,conf}}$, and ΔH_{conf} [27]). This was done for the V142G mutant and its experimental parameters were used to obtain fitted estimates for the other mutants, under the strong assumption that the conformational process was identical for all species.

Two important findings emerged from the calorimetric data. First, the enthalpy difference between the BI and BC states is significant: $\Delta H_{\mathrm{conf}} = 33 \pm 1$ kcal/mol at 35.1°C (Table 3). As described below, this value is almost exactly what would be expected for the enthalpy of unfolding of the entire LID domain. Second, the calorimetric data obtained for all Val to Gly mutants and wild-type could be reasonably fit with this common ΔH_{conf}, a common $\Delta C_{\mathrm{p,conf}}$ of 660 ± 70 cal/(mol K), and an individual $T_{\mathrm{m,conf}}$ (the midpoint temperature of the BC to BI transition). As can be seen in Table 3, the mutation effects on the local unfolding transition can be large, shifting from the $T_{\mathrm{m,conf}} = 52$°C of wild-type to 35°C for V142G.

The importance of these calorimetric results is threefold. First, the fact that only the $T_{\mathrm{m,conf}}$ differs between each mutant suggests that the mutations have indeed selectively increased the conformational entropy of the BI state, which might be thermodynamically considered a locally unfolded state. Second, the conformational and thermodynamic character of the BI state appears common to the wild-type

Table 3 Thermodynamic parameter estimates for AK(e) and mutants BI to BC conformational transitions obtained from mathematical fitting of ITC data

	$\Delta H_{m,conf}$ (kcal/mol)[a]	$\Delta C_{p,conf}$ (cal/(mol K))[b]	$T_{m,conf}$ (°C)[c]
V142G	33 ± 1	660 ± 60	35.1 ± 0.3
V148G	Unchanged from V142G[d]	Unchanged from V142G	37.9 ± 0.3
V135G	Unchanged from V142G	Unchanged from V142G	41.1 ± 0.2
Wild-type	Unchanged from V142G	Unchanged from V142G	52.3 ± 0.1

[a]Experimentally observed transition enthalpy at $T_{m,conf}$ of V142G
[b]Experimentally observed heat capacity of V142G
[c]Transition midpoint temperature, error is 95% confidence interval estimated from the fitted model
[d]"Unchanged from V142G" means that the V142G value was used in the fitting routine for this mutant. The fitting routine has been described in detail [27]

AK(e) and all three mutants. Third, the transition between the BI and BC states is consistent with a cooperative two-state process [27].

What is the relation between the binding reaction and the local unfolding event in the LID domain? The most reasonable answer is that they are thermodynamically indistinguishable: the BI state demonstrated by ITC could be the random coil state of the LID domain demonstrated by the globally fit CPMG results.

A quantitative link between the two states is provided by the COREX algorithm [17], which uses the high resolution structure as a template in conjunction with a long-standing surface area based parameterization [62] of protein energetics to predict the ΔH and ΔC_p of unfolding of different regions of the protein. If the BI state were indeed a locally unfolded state, a reasonable model based on the NMR data would be to assume that residues 110–164 are unfolded in the BI state (c.f. Fig. 6b). Using the COREX energy function, the predicted thermodynamics of unfolding this region in wild-type were compared with the experimentally determined values. The agreement between calculation and experiment is excellent (Table 4). The calculation for global unfolding of V142G also matches the experimental value, indicating that a two state approximation for global unfolding of the entire molecule, as used for the CD data in Fig. 5, was appropriate.

3.4 *Estimating the Populations in a Three State Model of AK(e) Conformational Equilibrium*

In summary, multiple lines of experimental (ITC, HSQC, CPMG, CD, crystallographic), computational (equilibrium simulation, COREX), and theoretical (gain-of-entropy) evidence are converging on the notion that the conformational change in the AK(e) LID domain is more complex than a simple rigid-body opening and closing. Specifically, the data demonstrate that the LID domain can transiently

Table 4 Experimental and predicted thermodynamics of local and global unfolding

		Local[a]		Global[a]		Sum[a]	
Parameter	T (°C)	Isothermal titration calorimetry[c]	COREX[d,e]	Circular dichroism[c]	COREX[e,f]	Experimentally determined[g]	COREX[e,h]
ΔH (kcal/mol)	35.1	33 ± 1[b]	32.7	34 ± 9[i]	40.6	67 ± 10[k]	73.3
	54.7	46 ± 2[j]	46.5	89 ± 4	88.2	135 ± 6[k]	134.6
ΔC_p (kcal/(mol K))		0.66 ± 0.6	0.7	2.8 ± 0.2	2.4	3.5 ± 0.3[k]	3.1

[a]Paired experimental and computational measures of possible two state unfolding reactions
[b]Errors represent experimental 95% confidence intervals
[c]Experimental parameter estimates from ITC (Fig. 9a, inset) and CD of V142G (Fig. 5), respectively
[d]Residues 110–164 of AK(e)
[e]Values computed using the COREX energy function and the PDB 4AKE structure, errors are assumed to be within ±10%.
[f]Residues 1–109 and 165–214 of AK(e)
[g]Sum of enthalpy and heat capacity for the BI to BC reaction (ITC) and the native to denatured transition (CD)
[h]Residues 1–214 of AK(e)
[i]Calculated as ΔH_{global} (54.7) + $\Delta C_{p,global}$ (35.1–54.7).
[j]Calculated as ΔH_{local} (35.1) + $\Delta C_{p,local}$ (54.7–35.1)
[k]Propagated 95% Confidence Interval.

unfold from a binding competent (BC) state to a binding incompetent state (BI), in equilibrium with a fully unfolded (U) state:

$$[\mathrm{BC} \rightleftarrows \mathrm{BI}] \rightleftarrows U \tag{9}$$

According to this formalism, the populations of all three states can be simulated [27] given K_{conf} (the equilibrium between BC and BI in (9), estimated from ITC data in Table 4) and K_{unf} (the equilibrium between globally folded and unfolded states, estimated from CD data in Table 4). The results indicate that the locally unfolded BI state is substantially populated for the wild-type enzyme to a level of about 5% at 37°C (Fig. 10, orange bar). For the V142G mutant, a similar population exists at a lower temperature, approximately 20°C (Fig. 10). Importantly, this latter value quantitatively agrees with the 3–9% populations obtained by the CPMG experiments as described above in Sect. 3.2. The fact that two independent biophysical techniques (calorimetry and NMR), which are each sensitive to different properties of the protein, yielded similar results strongly suggests that the model whereby the crystallographic "open" and "closed" conformations are in equilibrium is simply incorrect (or at best, incomplete).

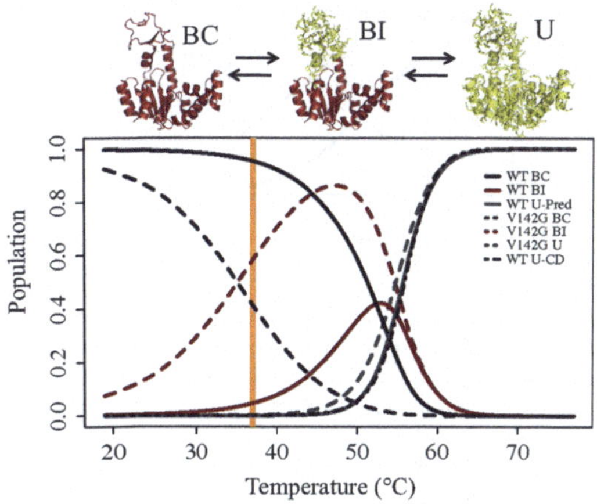

Fig. 10 Population simulations for wild-type AK(e) and V142G agree with experimentally determined CPMG population. In the molecular figures, *red* indicates folded conformation and *yellow* indicated unfolded (*U*) conformation. Populations are calculated from the BI–BC (K_{conf}) equilibrium estimated from the ITC experimental data, and from the U–BI ($K_{unfolded}$) equilibrium estimated from the CD unfolding experiments on V142G. (The validity of this approach can be appreciated in this figure from the high probability of the BI state at 47°C, before the global unfolding transition.) Thus, the partition function Q for this system is $Q = 1 + K_{conf} + (K_{conf} \times K_{unfolded})$. In this model, the probability of BC, BI, and U states are given by $1/Q$, K_{conf}/Q, and $K_{conf} \times K_{unfolded}/Q$, respectively. The *black dot-dashed line* for wild-type shows the population of the U state calculated directly from the CD unfolding data (Fig. 5, inset table). *Orange line* marks the optimal growth temperature of *E. coli*, 37°C

4 The AK(e) Functional Landscape: The Biologically Adaptive Potential of Local Unfolding of the LID Domain

4.1 The Population of the Unfolded LID Domain Regulates the K_m *of the Enzyme*

The data reviewed here require construction of an equilibrium model for the conformational states of the LID domain that is more complex than the interconversion between two ostensibly folded "open" and "closed" states. Is this new model, which predicts local unfolding of the LID, relevant to the function of AK (e)? The answer is of course yes. Because the population of BI state is 5% under physiological conditions, the equilibrium between the BI and BC states *tunes* the K_m of the enzyme, as an increased population of states that do not bind substrate will decrease the apparent affinity of enzyme for substrate. What are the implications for biological adaptation of AK and possibly other enzymes?

To adapt to a new environment an enzyme must alter its kinetic parameters such as K_m and k_{cat}, while simultaneously maintaining the structural properties of the catalytic site, which permit the functional chemistry and the molecular recognition

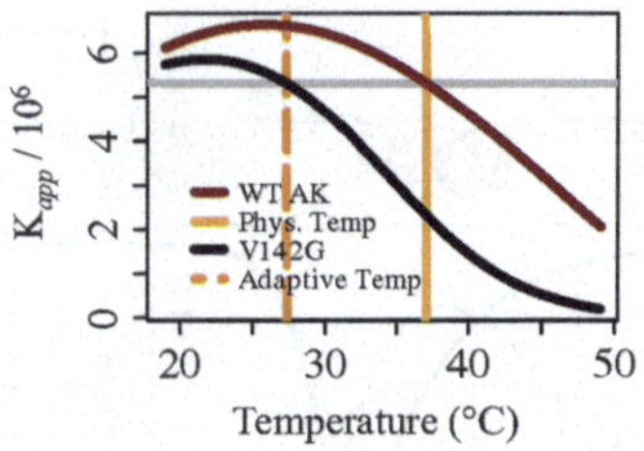

	T_{adapt} (°C)
v142g	27.3
v148g	29.5
v135g	30.0
WT	37.0

Fig. 11 Temperature adaptation of K_m arising from surface exposed glycine mutations that redistribute the ensemble populations of binding incompetent (BI) conformational states. Plot of the apparent binding affinity (K_{app}) for Ap5A, computed from fitted ITC data (Fig. 9). The *dashed orange line* marks the temperature where K_{app} of V142G is equivalent to K_{app} of wild-type AK(e) at 37°C. The *dashed orange line* can therefore be considered as a surrogate homeostatic temperature for V142G. In other words, V142G could be a cold-adapted enzyme due to the single glycine mutation. The *table* indicates the new surrogate homeostatic temperatures computed for the remaining mutants

of the substrate. Numerous studies have demonstrated that one common adaptive strategy that achieves this goal is to allow changes in regions of the protein distant to the active site, with the architecture and functional groups of the active site proper remaining highly conserved [12, 30, 63–65]. In the specific case of cold temperature adaptation, the main target of adaptive mutations appears to be increased conformational flexibility [8, 10]. Somero and colleagues have previously hypothesized that, in cold-adapted enzymes, decreased substrate affinity is a consequence of a more flexible native state exploring nonbinding conformations.

The experimental results presented in this chapter support that hypothesis and demonstrate how changes to the populations of low affinity or binding incompetent (BI) excited states, rather than perturbing the structure of the fully folded ground state, can effect adaptive change. Such potential adaptation can be visualized by plotting the temperature dependence of the fitted apparent association constant (K_{app}) of inhibitor binding for the wild-type AK(e) and V142G enzymes (Fig. 11). Considering K_{app} as a surrogate adaptive endpoint and 37°C as the original homeostatic temperature, the V142G mutation reduces the surrogate homeostatic temperature [i.e., $K_{app,mutant}(T_{adaptive}) = K_{app,wild\text{-}type}$ (37°C)] by approximately 10°C [27]. Thus, a single adaptive functional change can be mediated by a single point mutation. Of course, this phenomenon could apply to any protein–ligand binding reaction with similar thermodynamic features.

4.2 *The Biological Importance of Enzyme Functional Adaptability Through Local Unfolding*

The importance of this adaptation strategy is twofold. First, the change in binding occurs in the absence of active site changes or core structural deformation of the protein. Instead, the affinity is modulated by changing the conformational fluctuations around the protein's average structure. To date, structure-based protein design strategies have been largely based on static representations of the protein and ligand [5, 66]. In contrast, these experimental results reveal an alternative strategy for evolutionary adaptation or rational design that targets fluctuations rather than structure. The fact that all three valine to glycine mutations produced similar effects suggests that this approach is robust. Second, the increased population of binding incompetent states among the members of the native state ensemble, as proposed by Somero and colleagues [8], is directly confirmed to be an effective strategy for the adaptive modulation of binding affinity.

Somero and colleagues hypothesized that k_{cat} could also be affected by flexibility-altering mutations. Preliminary results suggest that, under substrate saturating conditions, the equilibrium between BC and BI forms of AK is biased towards the substrate-bound conformation. Under such conditions, the rate of interconversion between BC and BI, as measured by CPMG, is equivalent for both wild-type AK(e) and the V142G mutant (Table 2). Moreover, this rate of interconversion is quantitatively indistinguishable from (not faster than) k_{cat} [67]. This last result implies that the BC to BI local unfolding equilibrium is relevant not only to K_m but also to product release, a conclusion in agreement with previously published results from other investigators [11, 67]. If true, local unfolding of the lid domain, mediated by single surface exposed glycine mutations, would dominate the evolutionary adaptability of both the binding and catalytic function of this essential enzyme.

4.3 Perturbation of the Ensemble, in Addition to Inspection of Static Structure, Is Necessary to Understand the Functional Landscape

It cannot be overemphasized that the functional adaptability conferred by the local unfolding of the LID domain is a consequence of the ensemble nature of AK(e). In general, the ensemble nature of proteins means that the phenotypic impact of mutation on function or stability will depend on how the ensemble is distributed, or "poised", prior to the mutation [25, 26]. Of crucial importance is the realization that such insight is *impossible* to obtain from analysis of a single structure, such as a ground state X-ray structure. This can be conceptually demonstrated with a highly simplified protein ensemble consisting of only two states (Fig. 12). One state in this ensemble consists of the entire protein folded (i.e., a binding competent (BC) ground state structure), and the second state partially unfolds a subdomain (i.e., a binding incompetent (BI) excited state structure). For this thermodynamic system, the effect of a valine to glycine mutation (e.g., the surface exposed mutants discussed in this chapter) on the ability to bind ligand will depend on where the equilibrium between the two states is poised. Generally, the range of responses can be divided into three categories (red, yellow, and orange regions in Fig. 12); a detailed discussion has been previously published [68]. Briefly, in the red region,

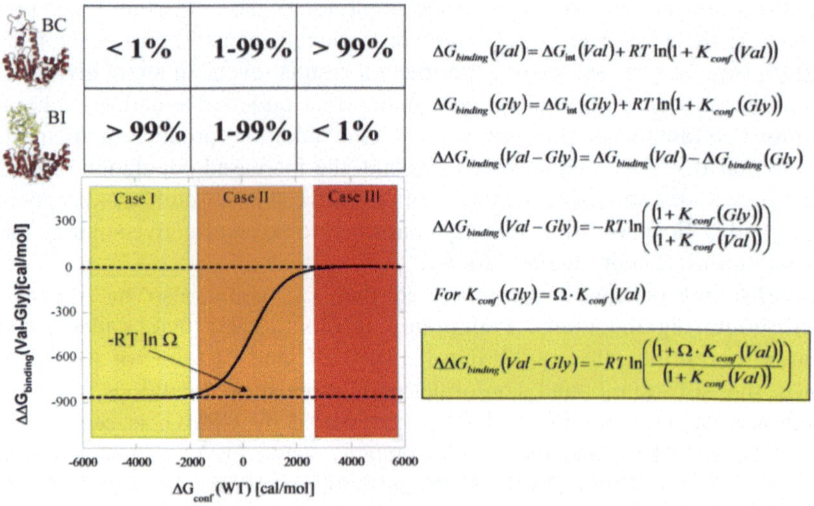

Fig. 12 Expected outcomes for Val to Gly mutation effects on ligand binding. The simplest case of a two-state ensemble is displayed. In the molecular picture, *red* indicates folded parts of the protein and *yellow* indicates unfolded parts. Note that the diagrams and equations are general and may apply to many enzyme–ligand systems, although they describe the specific case of AK(e) and its LID domain as discussed in the text. The equations are restatements and syntheses of equations previously discussed in the text: symbol Ω, denoting conformational degrees of freedom, was introduced in (1) and (2), K_{conf} and ΔG_{conf} in (7) and (8). Case II (*orange*) is the energetic region where the greatest experimentally observed changes would occur, and the approximately 5% unfolded LID domain for wild-type AK(e) places it in this region

where the BC state is favored, the mutation will have no effect. In contrast, in the yellow region, where the BI state is favored, the mutation will have a maximum effect. When both states are populated, the effect will be intermediate.

Closer analysis of these different cases reveals an important point. For any ensemble involving two or more states, there will always be linkage effects and thus a range of values (*y*-axis of Fig. 12) possible for $\Delta\Delta G_{\mathrm{binding}}$ (Val-Gly), depending on where the equilibrium is poised prior to the mutation. Consequently, a difference (or lack of a difference) in $\Delta\Delta G_{\mathrm{binding}}$ (Val-Gly) does not by itself give insight into either the nature of the fluctuations (i.e., whether they are rigid-body hinge motions or local unfolding) or any possible coupling that exists between different parts of the molecule. Insight can be gained, however, by perturbing the equilibrium and observing how the binding is affected (as was done for AK(e) in Fig. 9). Thus, a strategy to determine for any enzyme where the equilibrium is poised is mutational perturbation and observation of the amplification or suppression of functional effects. For AK(e), the model depicted in Fig. 12 is precisely the case for the three LID domain valine to glycine entropy promoting mutations. By changing the temperature, the BC to BI equilibrium, and thus the $\Delta\Delta G_{\mathrm{binding}}$ (Val-Gly), is shifted in a manner proscribed by the expressions in Fig. 12.

5 Conclusion: An Emerging Correspondence Between the Energetic and Functional Landscapes of AK(e) Mediated by Excited State Conformational Fluctuations of Local Unfolding

By employing a mutation strategy that affected the conformational entropy of local unfolding without affecting the average folded structure, insight has been gained into the energetic and functional landscapes of AK(e). Multiple experimental techniques, including X-ray crystallography (Fig. 4), NMR spectroscopy (Figs. 6 and 7), circular dichroism spectroscopy (Fig. 5), and isothermal titration calorimetry (Fig. 9) have been brought to bear on the question. All experimental results, supported by thermodynamic simulation (Figs. 10 and 11, Table 4) agree on two major conclusions: (1) the LID domain of AK(e) undergoes an all-or-none local unfolding reaction that affects substrate binding to the enzyme and (2) this locally infolded state of the LID domain is populated to a level of approximately 5% in the apo form of the enzyme.

The previously unappreciated experimental fact is that the unfolded LID domain is a dominant influence on the energetic and functional landscapes of AK(e). This holds important, possibly paradigm-changing, implications for the investigation and understanding of how the enzyme works. In particular, these results force a revision of the intuitively appealing canonical picture of the molecule undergoing rigid-body opening and closing hinge motions during its catalytic cycle. This canonical picture has been the most common framework for molecular dynamics simulation, as the assumption of two well-defined conformational end-states permits the most efficient use of limited computational sampling resources. However, the results presented here reveal a need for computational approaches that can thoroughly sample disorder (i.e., quantify conformational entropy on large systems), such as state of the art methods that can lengthen the timescales of atomic simulation [69].

Perhaps the most intriguing implications arise regarding the evolution of enzyme function and the adaptability of enzymes to different biological environments. As proposed many years ago by Somero [8] and others, flexibility can be key to temperature adaptation. The fact that single point mutants to glycine, in surface exposed locations of the enzyme distant from the active site, can conserve relative function at substantially different temperatures experimentally demonstrates a remarkably efficient mechanism for enzyme evolution. This mechanism may well be immediately applicable to the rational design of amino acid substitutions affecting function in other therapeutic or industrially important enzymes.

It is likely that detailed investigation of the amino acid substitution patterns in the LID domains of other AK homologs (Fig. 2) will permit rationalization of those patterns in energetic terms that cannot presently be explained by the general physico-chemical rules summarized within amino acid substitution matrices. Especially interesting would be a functional explanation of why the LID domain seems to be "binary": either present or absent in the biological record (Fig. 2). Perhaps the

LID domain was an evolutionary innovation, biophysically rooted in the energetic linkage described in Fig. 12, which allowed greater catalytic control over this metabolically indispensable enzyme.

References

1. Lehninger AL (1975) Biochemistry. Worth, New York
2. Creighton TL (1993) Proteins: structures and molecular properties. W.H. Freeman and Company, New York
3. Berg JM, Tymoczko JL, Stryer L (2007) Biochemistry. W.H. Freeman and Company, New York
4. Fersht AR (1998) Structure and mechanism in protein science: a guide to enzyme catalysis and protein binding. W.H. Freeman and Company, New York
5. Roethlisberger D, Khersonsky O, Wollacott AM, Jiang L, DeChancie J, Betker J, Gallaher JL, Althoff EA, Zanghellini A, Dym O, Albeck S, Houk KN, Tawfik DS, Baker D (2008) Nature 453:190
6. Michaelis L, Menten ML, Johnson KA, Goody RS (2011) Biochemistry 50:8264
7. Georlette D, Blaise V, Collins T, D'Amico S, Gratia E, Hoyoux A, Marx JC, Sonan G, Feller G, Gerday C (2004) FEMS Microbiol Rev 28:25
8. Fields PA, Somero GN (1998) Proc Natl Acad Sci USA 95:11476
9. Hammes GG, Benkovic SJ, Hammes-Schiffer S (2011) Biochemistry 50:10422
10. Zavodszky P, Kardos J, Svingor A, Petsko GA (1998) Proc Natl Acad Sci USA 95:7406
11. Wolf-Watz M, Thai V, Henzler-Wildman K, Hadjipavlou G, Eisenmesser EZ, Kern D (2004) Nat Struct Mol Biol 11:945
12. Henzler-Wildman K, Kern D (2007) Nature 450:964
13. Henzler-Wildman KA, Lei M, Thai V, Kerns SJ, Karplus M, Kern D (2007) Nature 450:913
14. Bosco DA, Eisenmesser EZ, Pochapsky S, Sundquist WI, Kern D (2002) Proc Natl Acad Sci USA 99:5247
15. Miyashita O, Onuchic JN, Wolynes PG (2003) Proc Natl Acad Sci USA 100:12570
16. Henzler-Wildman KA, Thai V, Lei M, Ott M, Wolf-Watz M, Fenn T, Pozharski E, Wilson MA, Petsko GA, Karplus M, Hubner CG, Kern D (2007) Nature 450:838
17. Hilser VJ, Garcia-Moreno EB, Oas TG, Kapp G, Whitten ST (2006) Chem Rev 106:1545
18. Hilser VJ, Freire E (1997) Proteins 27:171
19. Hilser VJ, Freire E (1996) J Mol Biol 262:756
20. Liu T, Pantazatos D, Li S, Hamuro Y, Hilser VJ, Woods VLJ (2012) J Am Soc Mass Spectrom 23:43
21. Pan H, Lee JC, Hilser VJ (2000) Proc Natl Acad Sci USA 97:12020
22. Motlagh H, Hilser VJ (2012) Proc Natl Acad Sci USA 109:4134
23. Whitten ST, Garcia-Moreno EB, Hilser VJ (2005) Proc Natl Acad Sci USA 102:4282
24. Wooll JO, Wrabl JO, Hilser VJ (2000) J Mol Biol 301:247
25. Hilser VJ, Thompson EB (2007) Proc Natl Acad Sci USA 104:8311
26. Hilser VJ, Wrabl JO, Motlagh H (2012) Annu Rev Biophys 41:585
27. Schrank TP, Bolen DW, Hilser VJ (2009) Proc Natl Acad Sci USA 106:16984
28. Schrank TP, Elam WA, Li J, Hilser VJ (2011) Methods Enzymol 492:253
29. Liu T, Whitten ST, Hilser VJ (2007) Proc Natl Acad Sci USA 104:4347
30. Feller G, Gerday C (2003) Nature reviews. Microbiology 1:200
31. Wrabl JO, Gu J, Liu T, Schrank TP, Whitten ST, Hilser VJ (2011) Biophys Chem 159:129
32. Aden J, Wolf-Watz M (2007) J Am Chem Soc 129:14003
33. Muller CW, Schlauderer GJ, Reinstein J, Schulz GE (1996) Structure 4:147
34. Muller CW, Schulz GE (1992) J Mol Biol 224:159

35. Sheng XR, Li X, Pan XM (1999) J Biol Chem 274:22238
36. Tsai MD, Yan HG (1991) Biochemistry 30:6806
37. Fukami-Kobayashi K, Nosaka M, Nakazawa A, Go M (1996) FEBS Lett 385:214
38. Schulz GE, Schiltz E, Tomasselli AG, Frank R, Brune M, Wittinghofer A, Schirmer RH (1986) Eur J Biochem 161:127
39. Hamada M, Sumida M, Kurokawa Y, Sunayashiki-Kusuzaki K, Okuda H, Watanabe T, Kuby SA (1985) J Biol Chem 260:11595
40. Cukier RI (2009) J Phys Chem B 113:1662
41. Maragakis P, Karplus M (2005) J Mol Biol 352:807
42. Daily MD, Phillips GN Jr, Cui Q (2010) J Mol Biol 400:618
43. Beckstein O, Denning EJ, Perilla JR, Woolf TB (2009) J Mol Biol 394:160
44. Chu JW, Voth GA (2007) Biophys J 93:3860
45. Arora K, Brooks CL 3rd (2007) Proc Natl Acad Sci USA 104:18496
46. Brylinski M, Skolnick J (2008) Proteins 70:363
47. Munier-Lehmann H, Burlacu-Miron S, Craescu CT, Mantsch HH, Schultz CP (1999) Proteins 36:238
48. Auton M, Bolen DW (2007) Methods Enzymol 428:397
49. D'Aquino JA, Gomez J, Hilser VJ, Lee KH, Amzel LM, Freire E (1996) Proteins 25:143
50. Creamer TP (2000) Proteins 40:443
51. Manson A, Whitten ST, Ferreon JC, Fox RO, Hilser VJ (2009) J Am Chem Soc 131:6785
52. Whitten ST, Yang HW, Fox RO, Hilser VJ (2008) Protein Sci 17:1200
53. Leach SJ, Nemethy G, Scheraga HA (1966) Biopolymers 4:369
54. Loria JP, Rance M, Palmer AG (1999) J Am Chem Soc 121:2331
55. Tollinger M, Skrynnikov NR, Mulder FA, Forman-Kay JD, Kay LE (2001) J Am Chem Soc 123:11341
56. Palmer AG 3rd, Kroenke CD, Loria JP (2001) Methods Enzymol 339:204
57. Kern D, Eisenmesser EZ, Wolf-Watz M (2005) Methods Enzymol 394:507
58. Vallurupalli P, Kay LE (2006) Proc Natl Acad Sci USA 103:11910
59. McConnell HM (1958) J Chem Phys 28:430
60. Braun D, Wider G, Wuethrich K (1994) J Am Chem Soc 116:8466
61. Eftink MR, Anusiem AC, Biltonen RL (1983) Biochemistry 22:3884
62. Freire E (2001) Methods Mol Biol 168:37
63. Russell NJ (2000) Extremophiles 4:83
64. Deng H, Zheng J, Clarke A, Holbrook JJ, Callender R, Burgner JW 2nd (1994) Biochemistry 33:2297
65. Aghajari N, Feller G, Gerday C, Haser R (1998) Structure 6:1503
66. Hardy JA, Wells JA (2004) Curr Opin Struct Biol 14:706
67. Aden J, Verna A, Schug A, Wolf-Watz M (2012) J Am Chem Soc 134:16562
68. Ferreon JC, Hamburger JB, Hilser VJ (2004) J Am Chem Soc 126:12774
69. Dror RO, Dirks RM, Grossman JP, Xu H, Shaw DE (2012) Annu Rev Biophys 41:429
70. Zhang Y, Skolnick J (2005) Nucleic Acids Res 33:2302
71. Berman HM, Westbrook J, Feng Z, Gilliland G, Bhat TN, Weissig H, Shindyalov IN, Bourne PE (2000) Nucleic Acids Res 28:235
72. Huang Y, Niu B, Gao Y, Fu L, Li W (2010) Bioinformatics 26:680
73. Pei J, Kim BH, Grishin NV (2008) Nucleic Acids Res 36:2295
74. Goodstadt L, Ponting CP (2001) Bioinformatics 17:845
75. Burlacu-Miron S, Gilles AM, Popescu A, Barzu O, Craescu CT (1999) Eur J Biochem 264:765
76. Ponder JW, Richards FM (1987) J Mol Biol 193:775
77. Ramachandran GN, Sasisekharan V (1968) Adv Protein Chem 23:283
78. Ramakrishnan C, Ramachandran GN (1965) Biophys J 5:909
79. Wuethrich K (1986) NMR of proteins and nucleic acids. Wiley, New York

Top Curr Chem (2013) 337: 123–138
DOI: 10.1007/128_2012_316

Published online: 10 April 2013

Structured Crowding and Its Effects on Enzyme Catalysis

Buyong Ma and Ruth Nussinov

Abstract Macromolecular crowding decreases the diffusion rate, shifts the equilibrium of protein–protein and protein–substrate interactions, and changes protein conformational dynamics. Collectively, these effects contribute to enzyme catalysis. Here we describe how crowding may bias the conformational change and dynamics of enzyme populations and in this way affect catalysis. Crowding effects have been studied using artificial crowding agents and in vivo-like environments. These studies revealed a correlation between protein dynamics and function in the crowded environment. We suggest that crowded environments be classified into *uniform crowding* and *structured crowding*. Uniform crowding represents random crowding conditions created by synthetic particles with a narrow size distribution. Structured crowding refers to the highly coordinated cellular environment, where proteins and other macromolecules are clustered and organized. In structured crowded environments the perturbation of protein thermal stability may be lower; however, it may still be able to modulate functions effectively and dynamically. Dynamic, allosteric enzymes could be more sensitive to cellular perturbations if their free energy landscape is flatter around the native state; on the other hand, if their free energy landscape is rougher, with high kinetic barriers separating deep minima, they could be more robust. Above all, cells are structured; and this holds both for the cytosol and for the membrane

B. Ma (✉)
Basic Science Program, SAIC-Frederick, Inc., Center for Cancer Research Nanobiology Program, Frederick National Laboratory for Cancer Research, National Cancer Institute, Frederick, MD 21702, USA
e-mail: mabuyong@mail.nih.gov

R. Nussinov (✉)
Basic Science Program, SAIC-Frederick, Inc., Center for Cancer Research Nanobiology Program, Frederick National Laboratory for Cancer Research, National Cancer Institute, Frederick, MD 21702, USA

Department of Human Genetics and Molecular Medicine, Sackler School of Medicine, Sackler Institute of Molecular Medicine, Tel Aviv University,
Tel Aviv 69978, Israel
e-mail: ruthnu@helix.nih.gov

environment. The crowded environment is organized, which limits the search, and the crowders are not necessarily inert. More likely, they too transmit allosteric effects, and as such play important functional roles. Overall, *structured cellular crowding* may lead to higher enzyme efficiency and specificity.

Keywords Allostery · Conformational selection · Energy landscape · Enzymatics · Macromolecular crowding · Protein dynamics

Contents

1 Introduction

Even though it is still highly debatable whether enzyme dynamic motions contribute to decreasing the chemical reaction barrier, the consensus picture emerging from experimental and computational studies indicates that enzyme conformational transitions are highly organized with stepwise conformational selection in catalysis, which increases enzyme specificity and efficiency [1, 2]. The close coupling of enzyme conformational dynamics and catalysis can be viewed as the outcome of the enzyme's adaptation to aqueous-based life. Stable, folded macromolecules in crystal structures have well defined three-dimensional structures, which are not as dynamic as those in aqueous solution. In solution, enzymes exist as conformational ensembles and their populations follow thermodynamic distributions [3–5]. Local energy fluctuations in water range between 10 and 20 kcal/mol [6], which is enough to perturb a well folded enzyme to a vast number of states. Enzyme dynamics in catalysis are not only real – they are inevitable [2].

Enzyme dynamics may involve small or large amplitude conformational changes. Large conformational change may involve nanoscale movements of domains or loops [7]. One example is the dengue virus NS2B-NS3 protease, for which the functionally important C-terminal segment of the NS2B cofactor dissociates in the open state via a large structural change to produce the closed state and confer activity [8] through a conformational selection mechanism [9]. Large conformational changes typically relate to function, for example, to accommodate the conformational reorganization required to stabilize the transition states [10] or to allow allosteric regulation [11, 12]. The enzyme's dynamic conformational changes are also necessary for signal transduction across long distances in the cell [7, 13].

The enzyme's native playground is in a highly crowded cellular environment. In the *Escherichia coli* cell, proteins and nucleic acids can occupy 20–30% of the total volume [14]. Macromolecular crowding decreases the diffusion rate [14, 15], shifts the equilibrium of protein–protein association and of protein–substrate interaction, and changes the protein's conformational dynamics. These effects collectively contribute to enzyme catalysis. Enzymes respond differently to crowding agents; the activity of some decreases and of others increases, and the extents differ as well. For example, while macromolecular crowding has minimal effects on the kinetics and function of yeast hexokinase in physiological solutions [16], it can significantly increase the thermal stability of catalase; and the overall structure becomes more rigid [17]. Crowding effects are also concentration-dependent. For the monomeric multi-copper oxidase, *Saccharomyces cerevisiae* Fet3p, at low amounts of crowding agent, both the K_m (substrate binding) and the K_{cat} (catalytic efficiency) increase, whereas at higher crowding levels both parameters decrease [18]. Enzyme conformations also change with crowding conditions. The functional properties of the tryptophan synthase α2β2 complex in the presence of the crowding agents dextran 70 and ficoll 70 indicate that the rates of the conformational transitions which are associated with catalysis and regulation are reduced, and an open and less catalytically active conformation is stabilized [19]. Given these complex scenarios, one important question is whether the crowing effects simply reduce the extent of the enzyme dynamics or whether enzyme catalysis can benefit from a crowded environment. In this chapter we review experimental data, focusing on the relationship between macromolecular crowding and conformational dynamics as they relate to enzyme catalysis in a crowded environment. Experimental data present different crowding outcomes: this could be because of different experimental conditions, type of molecular crowder, and the concentration of the enzyme and its substrates. While bearing in mind these possible caveats, the results may also reflect the different types of systems and cofactors, enzyme conformations, and chemical reactions. Overall, the emerging picture suggests coupling between conformational dynamics and structured cellular crowding which could increase the enzyme specificity.

2 Correlation Between Crowding Effects and Enzyme Conformational Dynamics and Activity

The effects of crowding on internal protein dynamics can be well illustrated by the HIV-1 protease. The distance between a pair of flaps in the HIV-1 protease can vary from 5 Å in the closed form to 22 Å in the open form. Molecular dynamics simulations indicated that flap opening is significantly suppressed in a highly crowded environment [20, 21]. It is reasonable to expect that the decrease in the open state and the reduction of the enzyme–ligand diffusion encounter rate are likely to slow the in vivo enzymatic activity [20]. Unfortunately there is no published experimental study which would allow a comparison with these theoretical predictions for the HIV-1 protease. However, experimental studies of the reverse proteolysis, the formation of peptide bond, show that crowding can enhance protease-catalyzed synthesis of model peptides [22]. Extensive studies of α-chymotrypsin also revealed a correlation between protein dynamics and crowding effects.

It is well established that conformational dynamics of α-chymotrypsin correlate with its activity. α-Chymotrypsin undergoes a reversible conformational change from an inactive chymotrypsinogen-like structure at high pH to an active conformation at neutral pH. Molecular dynamics simulations revealed that bovine (active) and rat (inactive) chymotrypsin explore different regions of the conformational space, indicating that conformational dynamics is more important than the sequence differences for activation [23]. Consistent with the kinetic parameters observed by fluorescence stopped-flow spectroscopy, targeted molecular dynamics simulations revealed multiple pathways for chymotrypsin activation [24]. Banerjee and Pal studied the correlation of the conformational dynamics at the active site of α-chymotrypsin with enzyme activity. Interestingly, they found that the highest catalytic efficiency of α-chymotrypsin is at 37 °C, the typical body temperature of homeothermal animals. Site selective fluorescence circular dichroism (FDCD) studies reveal that the conformational flexibility of the enzyme affects the structural perturbation at the active site. Consistently, the hydrodynamic diameter of the α-chymotrypsin decreases considerably with increasing temperature, indicating that the enzyme is more compact at higher temperatures [25].

Amide H/D exchange kinetics indicated that glycosylation can stabilize α-chymo trypsin and reduce its structural dynamics [26]. Enzyme catalysis kinetics indicated that glycosylation does not affect substrate binding (K_S) but decreases the rate of the catalytic steps. Therefore, for α-chymotrypsin, it is clear that the higher structural fluctuations decrease the catalytic rate [27]. With a dynamic-activity correlation, one would expect that macromolecular crowding should decrease the structural dynamics and the activity of α-chymotrypsin. However, it has been observed that crowding could either increase or decrease the catalytic rate of α-chymotrypsin, depending on the crowding molecules that are added to the solution. The addition of poly(ethylene glycol) (PEG) as crowding molecules increases the affinity of the enzyme for its substrate; however, this is followed by a decrease in the turnover number (K_{cat}) [28]. Chemical linkage of PEG to α-chymotrypsin may represent an "extreme" of a crowded environment. PEG conjugation has been shown to increase the α-chymotrypsin thermostability and to decrease the protein structural dynamics [29]. While the dehydration/hydration processes may also play a role, solvation is energetically coupled to the conformational changes of α-chymotrypsin [28]. Adding dextrans of various molecular weights to the reaction solutions was also tested. An increase in the dextran concentration decreases v_{max} and increases K_m. While the increase in K_m can be attributed to the slower protein diffusion rate due to crowding [30], the overall change in the catalytic rate is consistent with the rate using the PEG as the crowding molecules, indicating protein stabilization by dextran crowding. It was also found that the exclusion volume, not the size of the dextran, changes the catalytic rate of α-chymotrypsin. However, the different behavior was observed using much larger nanoparticles to represent the crowded environment. Tetraethylene glycol (TEG) functionalized gold nanoparticles with 2 nm core diameters (AuTEG) enhance the α-chymotrypsin (ChT) enzyme activity in a substrate-selective fashion [31]. While the hydrolysis of three substrates [*N*-succinyl-L-phenylalanine-*p*-nitroanilide (SPNA), *N*-glutaryl-phenylalanine-*p*-nitroanilide (GPNA), and *N*-benzoyl-tyrosine-*p*-nitroanilide (BTNA)] was not affected by the crowding, a marked increase in activity (K_{cat}/K_m) with the most hydrophobic substrate *N*-succinyl-alanine-alanine-proline-phenylalanine-*p*-nitroanilide (TP) was observed. Interestingly, high molecular weight PEG was shown to have similar effects as the functionalized gold nanoparticles, in contrast to the lower molecular weight PEG [31].

3 Allosteric Control and Crowding

The change in protein dynamics in a crowded environment reflects an intrinsic property of dynamic proteins: allostery [32]. Direct kinetic assays using isothermal calorimetry have shown that molecular crowding and allosteric activators affect pyruvate kinase kinetics in similar ways [33]. Allosteric regulation of enzyme activity is always accompanied by changes in protein dynamics [34]. Figure 1 shows two examples of allosteric enzymes. The first (Fig. 1a) is a 145 residue mini intein, which catalyzes protein splicing [35]. Inteins are phylogenetically diverse self-splicing proteins that are of great functional, evolutionary, biotechnological, and medical interest [36]. It has been shown that intein activity can be regulated by a loop (connecting two β-strands from the N- and C-terminal intein subdomains of the mini-intein) and by the V67L mutation [35]. Active intein has a longer loop (VR96DVE99TGE102–L404). While reducing the loop length to VR96DVE99L404 (with the 100-through-403 region deleted) still preserves the activity, further shortening to VR96L404 inactivates the enzyme. The V67L mutation similarly illustrates the dynamic nature of the allosteric regulation of this intein, which globally enhances splicing and related cleavage reactions. While this mutation in the mini intein causes little change in crystal structures, it significantly slows the hydrogen-exchange rates globally, indicating a shift to more stable conformations and narrowing the ensemble distribution [35, 37]. The enhanced activity, together with the reduced dynamics of intein, is similar to the crowding effects on α-chymotrypsin discussed in the last section.

Figure 1b shows the structure of another allosteric protein SARS-Co 3CL (3CLpro) peptidase, which is structurally and functionally similar to α-chymotrypsin. The 3CLpro peptidase activities increase in crowded environments [38]. One possible reason is that the active form of 3CLpro is a dimer [38]. Alternatively, dynamics change, rather than dimer formation, could underlie the crowding effects of 3CLpro's activity.

Crowding could shift the monomer–dimer equilibrium to favor the dimer formation [38, 39]. This is the case for RNase A, which forms 3D domain-swapped oligomers with novel enzymatic and biological activities [40]. However, crowding effects do not necessarily favor the association of proteins to form an active oligomer state. A counter-example can be found in the muscle glycogen phosphorylase b (Phb). Phb is a classical-like allosteric protein with T–R transition, with the active R state being

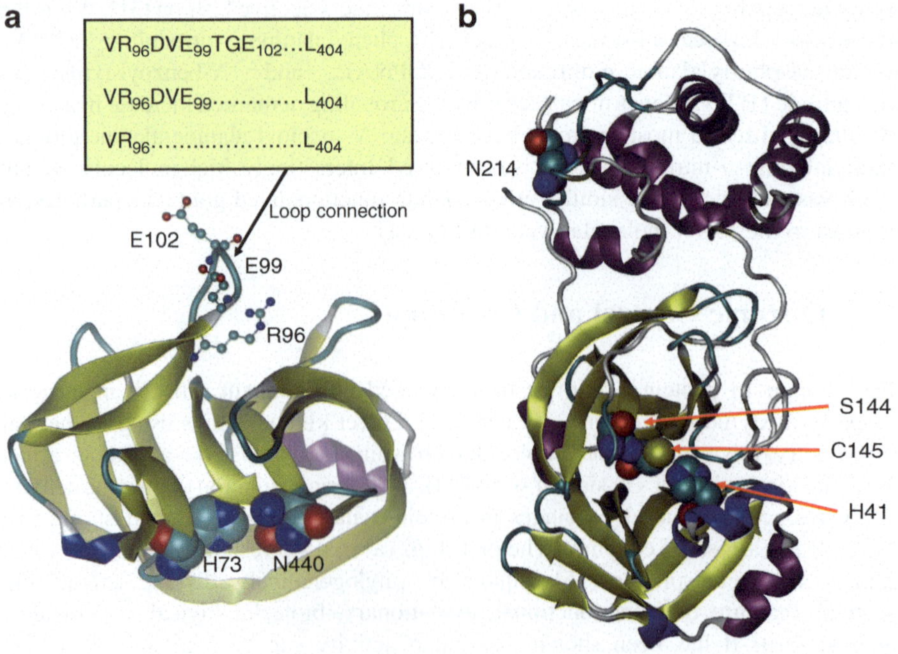

Fig. 1 Allosteric proteins are sensitive to crowding effects. (**a**) The mini intein activity can be allosterically modulated by a loop distant from the active site. Two residues in the active site (H73 and N440) are represented by *large balls* and residues in the loop are represented by *small balls* and *sticks*. Three loops with different sizes were tested in [35]. Active intein has a longer loop (VR96DVE99TGE102L404). While reducing the loop length to VR96DVE99. . .L404 still preserved the activity, further shortening to VR96L404 inactivates the enzyme. (**b**) The structure of allosteric protein SARS-Co 3CL peptidase, for which the catalytic activity increases in a crowded environment. Three active site residues (H41, S144, and C145) are shown as *large balls* in one domain. The mutant position N214 is shown as *large balls* in another domain

the smaller dimer and the inactive T state the tetramer. Adding osmolytes as crowding molecules favors the inactive unassociated T state [41, 42]. The reason could be that the dimer form of Phb is more compact. It is interesting to see that, under the same crowding conditions, the association of phosphorylase kinase (PhK) and its interaction with glycogen and the heat shock protein Hsp27 can be highly stimulated [43, 44].

On its own, the dimer form does not guarantee the activity of the 3CLpro. An inactive mutant with the N214A mutation in the extracellular domain of the 3CLpro still adopts a dimer structure which is almost identical to that of the wild-type [45]. The mutation site is distant from the active site. Molecular dynamics (MD) simulations revealed that the N214A mutant has much higher conformational flexibility than the WT enzyme [45]. Combining the crowding effects and the N214A mutational effects, one can see that a crowded environment restricts the conformational flexibility of the 3CLpro dimer; and thus increases its activity.

Isochorismate synthase (EntC) is another example which illustrates that macromolecular crowding increases the intrinsic activity of an enzyme by inducing conformational changes in the enzyme rather than through macromolecular association [46]. The EntC is a monomeric enzyme which catalyzes the reversible conversion between chorismate and isochorismate in *E. coli*. Ficoll addition leads to twofold increase of K_{cat}/K_m. In comparison to the EntC-catalyzed reaction, other reactions catalyzed by two homologous trimer and tetramer enzymes (EntB and LDH) are similarly stimulated by the same crowding conditions, indicating that protein–protein association is not a controlling factor for these enzymes. Changes of CD spectrum and Trp fluorescence revealed that conformational changes are responsible for the increase of the intrinsic activities of these enzymes [46].

Proteins can accommodate a crowded environment by different conformational changes. While red shift in Trp fluorescence was observed for EntC in crowded environments [46], blue shift was observed for the allosteric ADP-glucose pyrophosphorylase [47]. *E. coli* ADP-sugar pyrophosphatase (AspP) is an ultrasensitive hydrolase that catalyzes the hydrolytic breakdown of ADP-glucose linked to glycogen biosynthesis. The AspP activity is strongly enhanced by macromolecular crowding [48]. Molecular crowding renders AGPase more sensitive to the interplay between the allosteric regulators and consequently enhances the ultrasensitive response. Fourth-derivative spectroscopy and size-exclusion chromatography indicated that the ultrasensitive behavior is correlated with intramolecular conformational changes induced in the tertiary structure of the homo-tetrameric enzyme [47].

One of the common features of allosteric proteins is that their dynamics is designed to facilitate conformational change [49]. At the same time, from the functional standpoint, high sensitivity to crowding can present problems. The LRP protein is an example which illustrates that crowding effects could be coupled with function. The LRP transcriptional regulators are widely distributed among prokaryotes, bacteria, and archaea. The architecture of LRP proteins includes two distinct domains that harbor the regulatory (effector-binding) site and the active (DNA-binding) site, and the two domains are connected by a flexible hinge region. This structural feature and experiments suggest an allosteric switch for the LRP-like regulators [50–52]. LRP binding with DNA is stimulated under macromolecular crowding conditions [53]. In comparison, another DNA binding protein fis has no noticeable response to the same crowding conditions, indicating synergism of allostery and crowding effects for LRP DNA binding [53].

DNA binding proteins are similarly allosteric [54], and experimental data present mixed results with regard to the crowding effects. The hydrolysis of a 29-mer double-stranded DNA by DNase I and S1 nuclease was substantially enhanced by molecular crowding using PEG; however, molecular crowding had little effect on hydrolysis by exo III and exo I exonucleases, suggesting differential crowding effects on the catalytic activities under these conditions [55].

DNA replication is under high specificity pressure to ensure fidelity and retain efficiency under crowding conditions [56–60]. Crowding can help regulated stress response [61]. Under crowded conditions, dramatic improvements in all parameters of RT-PCR were observed, including eight- to tenfold greater sensitivity, enhanced polymerase processivity, higher specific amplicon yield, greater primer annealing and specificity, and enhanced DNA polymerase thermal stability [62]. Similarly, polymerase activity assays under various crowding conditions demonstrated that the activities of T7 and Taq DNA polymerases depend on the molecular weight and concentration of the crowding agent [63].

Inter- and intra-domain motions in well-defined regions of non-ribosomal peptide synthetases (NRPS) are important for its activity [64]. One of the NPRS, enterobactin synthetase, experiences significant conformational change in crowded solutions mimicking the intracellular environment. The structural change correlates well with the extent of the crowding-induced side product suppression in the nonribosomal enterobactin synthesis [65].

4 Enzyme Dynamics and Energy Landscapes with Artificial Uniform and In Vivo Structured Crowding

The free energy landscapes of protein folding and binding provide insights into protein structure and function which relate to allosteric regulation and enzyme catalysis [1, 2, 32, 66–68]. Macromolecular crowding changes the landscape, as evidenced by computational [69–71] and experimental studies [72, 73]. However, there are differences between the crowding effects caused by the artificial crowding under random conditions and the in vivo cellular environment. For example, there are differences (1) between proteins and artificial crowding agents and (2) between many copies of few proteins and the proteome. The corresponding free energy landscapes could differ substantially. Here we consider two kinds of crowding effects on the protein free energy landscape which represent two extreme cases. The first is uniform crowding and the second is structured crowding (Fig. 2). The question is how the free energy landscape, here represented by two conformations (α and β, Fig. 2a), could be affected.

Uniform crowding (Fig. 2b) considers effects caused by hard sphere crowding agents, which have uniform size and repulsive interactions with the protein solute. This scenario represents strong protein confinement. Crowding agents with sizes smaller than the protein solute could have larger effects. When the size of the crowding agent increases, there is an upper limit for the crowding effects, which has been shown by Milklos et al. [74]. Milklos et al. studied the stability and ps–ns internal dynamics of a small globular protein (radius ~2 nm) crowded by a large synthetic microgel particle (radii of ~300 nm). They found no change in protein rotational or ps–ns backbone dynamics; only mild stabilization at a volume occupancy of 70%, which approaches the occupancy of closely packed spheres [74]. However, it is difficult to predict the crowding effect when the size of the crowding agent is uniformly distributed. For example, proteins with aspherical shape may experience large conformational change in the presence of synthetic Ficoll70 [75]. Similarly, using Ficoll70 as crowding agent, a recent study with combined experiment and computer simulation demonstrated that macromolecular crowding dramatically affects the structure, function, and folding landscape of phosphoglycerate kinase (PGK) [73]. Overall, the relative stabilities among different conformers could either increase or decrease by uniform crowding (Fig. 2b).

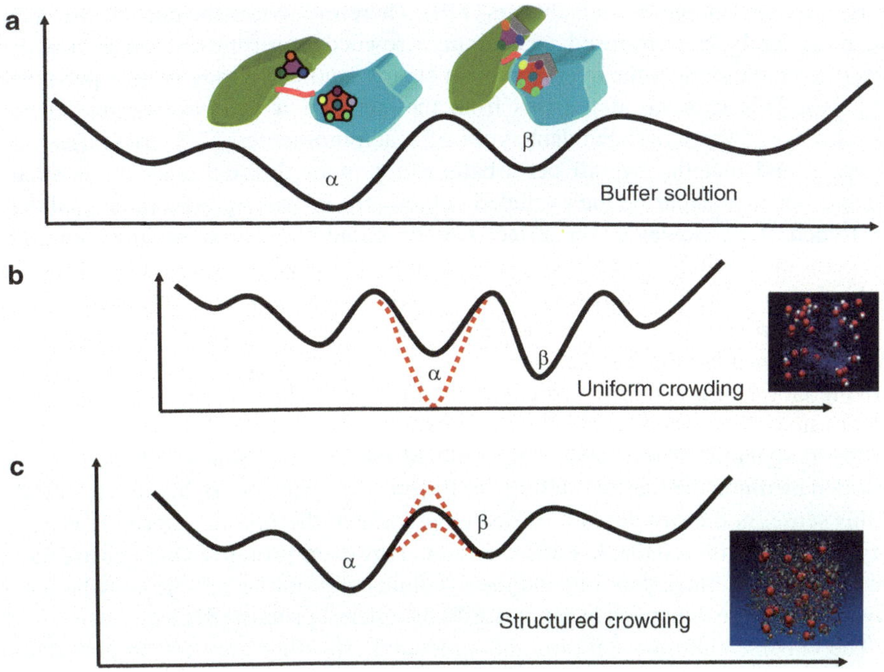

Fig. 2 Changes of the protein free energy landscape under uniform crowding and structured crowding environment. (**a**) The original energy landscape is represented by two conformational states α and β in a buffer solution. (**b**) Uniform crowding considers the effects caused by hard sphere crowding agents, which have uniform size and repulsive interactions with protein solute. This represents the strong confinement of proteins. In an environment of uniform crowding, the protein energy landscape may experience strong compression which may reflect protein folding–unfolding transition. Depending on the molecular size of the crowding agents, the relative stabilities of different conformers could increase or decrease in a uniform crowded environment. (**c**) Structured crowding refers to a highly coordinated cellular environment, where the overall proteome and other macromolecules pre-organized into structured clusters. In a structured crowded environment, the protein free energy landscape may be more similar to that in buffer solution. However, protein dynamics could be more sensitive than thermal stability

Structured crowding refers to the highly coordinated cellular environment, where the proteome and other macromolecules are pre-organized. Proteins often cluster and pre-organize to perform functions, for example, by forming transcriptional factories [76], which increases the effective local concentration. Pre-organization goes beyond clustering because it further reduces the diffusion-and-collision times: it positions the proteins in an orientation which is closer to the native functional one. Signal transduction occurs in an organized microenvironment [7, 13] where elements of a signaling pathway are connected functionally and spatially [77, 78]. The organization of proteins with different charges and sizes can be important for fast cellular response in fluctuating environments. There are hundreds of proteins in the cytoplasm, with different sizes and shapes. Our study of 206 kinds of proteins in a theoretical minimal proteome model indicated that proteins with different sizes and charges are well organized into clusters [110]. Therefore, while the particle diffusion can be severely compromised in a highly crowded environment, in a crowded *structured* environment the overall protein energy landscape may be less perturbed (Fig. 2c). This scenario also arises from the study of a cytoplasmic model that includes 50 of the most abundant types of macromolecules [79]. McGuffee and Elcock found that the overall perturbation of protein thermal stability is small, because when considering the excluded volume, the crowding effect may stabilize the folded state; however, the effect can be counterbalanced by the favorable energetic interactions which take place in unfolded conformations [79]. Proteins are better crowding agents than synthetic particles. For example, using two globular proteins (bovine serum albumin and hen-egg-white lysozyme) as crowding agents, chymotrypsin inhibitor 2 (CI2) was found to be only slightly destabilized [72]. Moving closer to a real crowded environment in the cell, hen egg white has been used to study the dynamics and stability of several proteins [80]. It was found that, while the dynamic parameters of the studied protein are clearly affected by the crowded medium, the thermal stability of the protein is similar to that in buffer [80]. In this sense, in the structured crowded environment, the protein energy landscape may be more similar to that in buffer solution. However, protein dynamics are more sensitive than thermal stability, and the dynamics of globular protein may be more sensitive than the dynamics of intrinsically disordered protein [81, 82].

The classification into uniform and structured crowding may be simplified and realistic crowding effects could be between these two ends of the spectrum. Protein dynamics may contribute more to the change in enzyme catalysis in crowded environments than thermal stability. This is well illustrated by the effect of a crowded solution on the reaction rates of the decarboxylating enzymes urease, pyruvate decarboxylase and glutamate decarboxylase, with the crowding agent being proteins and synthetic polymers [83]. It was found that increasing the concentration of globular proteins up to 30% crowding concentration caused a dramatic rise in enzyme activity; however, then the activity decreased; on the other hand, polymers caused a concentration-dependent decrease in activity [83].

5 Overview and Conclusions

Cells are crowded; and as John Ellis pointed out over a decade ago, while these effects are obvious they are (often) underappreciated [84]. For the *E. coli*, it was estimated that the cytosol contains about 300–400 mg/mL of macromolecules [85]. The bacterium contains up to 4,288 different types of proteins, with ~1,000 of these types produced at sufficiently high levels to be easily detected, DNA and RNA molecules [86]. Crowding reduces the volume of solvent available for other molecules in solution, thus increasing their effective concentrations. In eukaryotes, the interior of the cell is even more crowded: cells also contain protein filaments that make up the cytoskeleton. Pielak has highlighted crowding effects [87], and emphasized that experiments which are carried out on dilute samples (under10g/L) are very different to those in the living cell. This high crowding led Allen Minton to ask in his 2006 commentary [88] "How can biochemical reactions within cells differ from those in test tubes?" Minton argued that nonspecific interactions in the interior of the cell can greatly influence the equilibriums and rates of reactions. Minton classified the consequences of nonspecific interactions in the cellular interior into three types of phenomena: macromolecular crowding, confinement, and adsorption. He further pointed out that these nonspecific "background" interactions may be repulsive, which would lead to exclusion, and thus increase the effective concentration and chemical activity, or attractive, which would lead to nonspecific associations or adsorption. The predominantly repulsive background interactions may alter the dissociation constants by enhancing the rate and extent of macromolecular associations in solution [89]. Crowding may also affect enzyme reactions involving small molecules if there is a large change in the shape of the enzyme [90]. Reaction dynamics under crowded conditions are affected by a high concentration of reactants [91].

However, cells are not only crowded – they are also structured, and this holds not only for the membrane environment but for the cytoplasm as well [92, 93]. Hence, it is not only the membrane-bound compartmentalization and biochemical association with subcellular organelles; the structured macromolecular crowding also relates to the spatial regulation inside the cell. The microtubule-dependent organization of non-membranous components helps direct cellular function. The sequestration of GEF-H1 provides a clear example [94]. Thus, in contrast to the view of the cytoplasm as a free and fluid environment, the microtubule network provides a platform for a "structured cytoplasm". Binding to microtubules can be either directly or indirectly through other microtubule-binding scaffolding molecules – all dynamic and allosteric. This can cause partial sequestration of proteins by the microtubule network. The extent of sequestration can depend on binding affinities and microtubule density. The cytoskeleton mesh also effectively divides the cytosol into a network of narrow pores. Further, while highly dynamic, the genome is also pre-organized and highly structured. Structuring provides docking surfaces and is critical for function. The so-called matrix or scaffolding proteins which organize the spatial localization of protein assemblies have also been demonstrated to be hubs for controlling function as in the case of Ste5 in the MAPK signaling pathway [95, 96] and cullin [97]. Matrix proteins are also dynamic and allosterically regulate function [34, 98]. Thus, in vivo, proteins do not diffuse freely, and do not travel over long distances. Spatially organized crowded environment is a more realistic description; it also implies that allosteric effects can be transmitted via crowders. Protein crowders should not be viewed as necessarily inert media or of specific uniform shapes. During (specific or nonspecific) binding both partners can change their shapes or dynamics, which has been viewed as a "molecular dance" [99]. Further, residence times and molecular motility of the players involved are also important in the organization of the intracellular space. Overall, both cooperative protein–protein interactions and membrane properties play key roles in cell signaling and vesicle trafficking and thus in diffusion in the crowded cell. They allow switching between different protein states, either soluble in the cytoplasm or bound to the membrane, for example, by modulating the chemical composition of the membrane [100], by changing its physical properties such as curvature or fluidity, or by lipophilic post-translational modification such as myristoylation, palmitoylation, prenylation, etc. which are often allosteric. For the *E. coli* Min system, in vitro studies clarified the order of events in space and time and the interplay between cooperative binding of MinD to the membrane and positive feedback during protein detachment [101]. These in vivo considerations limit the space, distance to be travelled, and thus diffusion time in the spatially pre-organized and structured cell and may buffer conformational dynamic consequences.

To conclude, here we have focused on macromolecular crowding effects on enzyme conformations, dynamics, and catalysis. Computational and experimental studies revealed that enzyme conformations, dynamics, and catalytic rates differ under crowding conditions. Most of the experimental studies use synthetic crowding agents to probe crowding effects, which may not be realistic and may have only limited relevance to in vivo conditions [102, 103]; however, some in vivo-like conditions have also been published [79, 80, 82, 83]. Nevertheless, no matter what kinds of crowding agents were used, all provided valuable insight into the dynamic nature of proteins and protein functional adaptation to crowded environments. We suggest that crowded environments can be classified into uniform and structured crowding. Uniform crowding represents crowded conditions created by synthetic particles with narrow size distribution. Such conditions may elicit large protein structure and dynamic perturbations, and facilitate experimental investigation. Structured crowding refers to the highly coordinated cellular environment, where protein and other macromolecules are clustered and organized. Structured crowded environments may perturb the protein thermal stability less and allow dynamic modulation of protein function. Allosteric proteins with flat free energy landscape may be more sensitive to the crowded cellular environment. Proteins with rougher landscapes may be more robust. Dynamic regulation of allosteric enzymes may result in efficiency and specificity under crowding conditions.

While most studies focus on the crowded cytoplasm environment, it behoves us to note that cell membranes are also crowded. For example, the enzymatic activity of a proteolytic enzyme, the Subtilisin Carlsberg (SC) in anionic sodium dodecyl sulfate (SDS) micellar medium, has been explored and found to be retarded compared to that in bulk buffer [104]. *E. coli* dihydrofolate reductase (DHFR) is also a well studied allosteric enzyme [105–107], which also experiences large magnitudes of crowding effects on the vesicle surface [108]. Finally, similar to protein, RNA enzymes are also subjected to crowding effects. For example, the hammerhead ribozyme activity can be increased by a factor of two to six by PEG [109]. In particular, membrane environments are well known to be pre-organized. Thus, the underlying principle of coupling enzyme dynamics to activity in structured crowded environments is expected to be general.

Acknowledgments This project has been funded in whole or in part with Federal funds from the National Cancer Institute, National Institutes of Health, under contract number HHSN261200800001E. The content of this publication does not necessarily reflect the views or policies of the Department of Health and Human Services, nor does mention of trade names, commercial products, or organizations imply endorsement by the U.S. Government. This research was supported (in part) by the Intramural Research Program of the NIH, NCI, Center for Cancer Research.

References

1. Ma B, Nussinov R (2010) Curr Opin Chem Biol 14:652
2. Ma B, Kumar S, Tsai CJ, Hu Z, Nussinov R (2000) J Theor Biol 203:383
3. Ansari A, Berendzen J, Bowne SF, Frauenfelder H, Iben IET, Sauke TB, Shyamsunder E, Young RD (1985) Proc Natl Acad Sci USA 82:5000
4. Miller DW, Dill KA (1997) Protein Sci 6:2166
5. Dill KA, Chan HS (1997) Nat Struct Biol 4:10
6. Ohmine I, Tanaka H, Wolynes PG (1988) J Chem Phys 89:5852
7. Ma B, Nussinov R (2009) Proc Natl Acad Sci USA 106:6887
8. de la Cruz L, Nguyen TH, Ozawa K, Shin J, Graham B, Huber T, Otting G (2011) J Am Chem Soc 133:19205
9. Ekonomiuk D, Caflisch A (2009) Protein Sci 18:1003
10. Smith AJ, Muller R, Toscano MD, Kast P, Hellinga HW, Hilvert D, Houk KN (2008) J Am Chem Soc 130:15361
11. Osmulski PA, Hochstrasser M, Gaczynska M (2009) Structure 17:1137
12. Hindie V, Stroba A, Zhang H, Lopez-Garcia LA, Idrissova L, Zeuzem S, Hirschberg D, Schaeffer F, Jorgensen TJ, Engel M, Alzari PM, Biondi RM (2009) Nat Chem Biol 5:758
13. Nussinov R (2012) Mol Biosyst 8:22
14. Konopka MC, Shkel IA, Cayley S, Record MT, Weisshaar JC (2006) J Bacteriol 188:6115
15. Konopka MC, Sochacki KA, Bratton BP, Shkel IA, Record MT, Weisshaar JC (2009) J Bacteriol 191:231
16. Olsen SN, Ramlov H, Westh P (2007) Comp Biochem Physiol A Mol Integr Physiol 148:339
17. Belluzo S, Boeris V, Farruggia B, Pico G (2011) Int J Biol Macromol 49:936
18. Pozdnyakova I, Wittung-Stafshede P (2010) Biochim Biophys Acta 1804:740
19. Pioselli B, Bettati S, Mozzarelli A (2005) FEBS Lett 579:2197
20. Minh DD, Chang CE, Trylska J, Tozzini V, McCammon JA (2006) J Am Chem Soc 128:6006
21. Qin S, Minh DD, McCammon JA, Zhou HX (2010) J Phys Chem Lett 1:107
22. Somalinga BR, Roy RP (2002) J Biol Chem 277:43253
23. Matrai J, Verheyden G, Kruger P, Engelborghs Y (2004) Protein Sci 13:3139
24. Matrai J, Jonckheer A, Joris E, Kruger P, Carpenter E, Tuszynski J, De Maeyer M, Engelborghs Y (2008) Eur Biophys J 38:13
25. Banerjee D, Pal SK (2008) Langmuir 24:8163
26. Sola RJ, Griebenow K (2006) FEBS Lett 580:1685
27. Sola RJ, Griebenow K (2006) FEBS J 273:5303
28. Verma PK, Rakshit S, Mitra RK, Pal SK (2011) Biochimie 93:1424
29. Rodriguez-Martinez JA, Sola RJ, Castillo B, Cintron-Colon HR, Rivera-Rivera I, Barletta G, Griebenow K (2008) Biotechnol Bioeng 101:1142
30. Pastor I, Vilaseca E, Madurga S, Garces JL, Cascante M, Mas F (2011) J Phys Chem B 115: 1115
31. Jordan BJ, Hong R, Han G, Rana S, Rotello VM (2009) Nanotechnology 20:434004
32. Gunasekaran K, Ma B, Nussinov R (2004) Proteins 57:433
33. Lonhienne TG, Winzor DJ (2002) Biochemistry 41:6897
34. del Sol A, Tsai CJ, Ma B, Nussinov R (2009) Structure 17:1042
35. Hiraga K, Derbyshire V, Dansereau JT, Van Roey P, Belfort M (2005) J Mol Biol 354:916
36. Hiraga K, Soga I, Dansereau JT, Pereira B, Derbyshire V, Du Z, Wang C, Van Roey P, Belfort G, Belfort M (2009) J Mol Biol 393:1106
37. Du Z, Liu Y, Ban D, Lopez MM, Belfort M, Wang C (2010) J Mol Biol 400:755
38. Okamoto DN, Oliveira LC, Kondo MY, Cezari MH, Szeltner Z, Juhasz T, Juliano MA, Polgar L, Juliano L, Gouvea IE (2010) Biol Chem 391:1461
39. Wang W, Xu WX, Levy Y, Trizac E, Wolynes PG (2009) Proc Natl Acad Sci USA 106:5517
40. Ercole C, Lopez-Alonso JP, Font J, Ribo M, Vilanova M, Picone D, Laurents DV (2011) Arch Biochem Biophys 506:123

41. Chebotareva NA, Harding SE, Winzor DJ (2001) Eur J Biochem 268:506
42. Chebotareva NA, Kurganov BI, Harding SE, Winzor DJ (2005) Biophys Chem 113:61
43. Chebotareva NA (2007) Biochemistry (Mosc) 72:1478
44. Chebotareva NA, Makeeva VF, Bazhina SG, Eronina TB, Gusev NB, Kurganov BI (2010) Macromol Biosci 10:783
45. Shi J, Han N, Lim L, Lua S, Sivaraman J, Wang L, Mu Y, Song J (2011) PLoS Comput Biol 7:e1001084
46. Jiang M, Guo Z (2007) J Am Chem Soc 129:730
47. Casati DF, Aon MA, Iglesias AA (2000) Biochem J 350(Pt 1):139
48. Moran-Zorzano MT, Viale AM, Munoz FJ, Alonso-Casajus N, Eydallin GG, Zugasti B, Baroja-Fernandez E, Pozueta-Romero J (2007) FEBS Lett 581:1035
49. Li W, Wolynes PG, Takada S (2011) Proc Natl Acad Sci USA 108:3504
50. Brinkman AB, Ettema TJ, de Vos WM, van der Oost J (2003) Mol Microbiol 48:287
51. Ma B, Tsai CJ, Haliloglu T, Nussinov R (2011) Structure 19:907
52. Chen S, Iannolo M, Calvo JM (2005) J Mol Biol 345:251
53. Pul U, Wurm R, Wagner R (2007) J Mol Biol 366:900
54. Ma B, Tsai CJ, Pan Y, Nussinov R (2010) ACS Chem Biol 5:265
55. Sasaki Y, Miyoshi D, Sugimoto N (2007) Nucleic Acids Res 35:4086
56. Santoso Y, Joyce CM, Potapova O, Le Reste L, Hohlbein J, Torella JP, Grindley ND, Kapanidis AN (2010) Proc Natl Acad Sci USA 107:715
57. Andrade P, Martin MJ, Juarez R, Lopez de Saro F, Blanco L (2009) Proc Natl Acad Sci USA 106:16203
58. Eoff RL, Sanchez-Ponce R, Guengerich FP (2009) J Biol Chem 284:21090
59. Xu C, Maxwell BA, Brown JA, Zhang L, Suo Z (2009) PLoS Biol 7:e1000225
60. Rothwell PJ, Waksman G (2007) J Biol Chem 282:28884
61. Simon SM, Sousa FJ, Mohana-Borges R, Walker GC (2008) Proc Natl Acad Sci USA 105: 1152
62. Lareu RR, Harve KS, Raghunath M (2007) Biochem Biophys Res Commun 363:171
63. Sasaki Y, Miyoshi D, Sugimoto N (2006) Biotechnol J 1:440
64. Frueh DP, Arthanari H, Koglin A, Vosburg DA, Bennett AE, Walsh CT, Wagner G (2008) Nature 454:903
65. Guo ZF, Jiang M, Zheng S, Guo Z (2010) Bioorg Med Chem Lett 20:3855
66. Ma B, Kumar S, Tsai CJ, Nussinov R (1999) Protein Eng 12:713
67. Kumar S, Ma B, Tsai CJ, Sinha N, Nussinov R (2000) Protein Sci 9:10
68. Ma B, Shatsky M, Wolfson HJ, Nussinov R (2002) Protein Sci 11:184
69. Dong H, Qin S, Zhou HX (2010) PLoS Comput Biol 6:e1000833
70. Cheung MS, Klimov D, Thirumalai D (2005) Proc Natl Acad Sci USA 102:4753
71. Friedel M, Shea JE (2004) J Chem Phys 120:5809
72. Miklos AC, Sarkar M, Wang Y, Pielak GJ (2011) J Am Chem Soc 133:7116
73. Dhar A, Samiotakis A, Ebbinghaus S, Nienhaus L, Homouz D, Gruebele M, Cheung MS (2010) Proc Natl Acad Sci USA 107:17586
74. Miklos AC, Li C, Sorrell CD, Lyon LA, Pielak GJ (2011) BMC Biophys 4:13
75. Homouz D, Perham M, Samiotakis A, Cheung MS, Wittung-Stafshede P (2008) Proc Natl Acad Sci USA 105:11754
76. Bartlett J, Blagojevic J, Carter D, Eskiw C, Fromaget M, Job C, Shamsher M, Trindade IF, Xu M, Cook PR (2006) Biochem Soc Symp 73:67–75
77. Reth M, Wienands J (1997) Annu Rev Immunol 15:453
78. Maudsley S, Martin B, Luttrell LM (2005) J Pharmacol Exp Ther 314:485
79. McGuffee SR, Elcock AH (2010) PLoS Comput Biol 6:e1000694
80. Martorell G, Adrover M, Kelly G, Temussi PA, Pastore A (2011) Proteins 79:1408
81. Szasz CS, Alexa A, Toth K, Rakacs M, Langowski J, Tompa P (2011) Biochemistry 50:5834
82. Li C, Charlton LM, Lakkavaram A, Seagle C, Wang G, Young GB, Macdonald JM, Pielak GJ (2008) J Am Chem Soc 130:6310

83. Derham BK, Harding JJ (2006) Biochim Biophys Acta 1764:1000
84. Ellis RJ (2001) Trends Biochem Sci 26:597
85. Zimmerman SB, Trach SO (1991) J Mol Biol 222:599
86. Blattner FR, Plunkett G 3rd, Bloch CA, Perna NT, Burland V, Riley M, Collado-Vides J, Glasner JD, Rode CK, Mayhew GF, Gregor J, Davis NW, Kirkpatrick HA, Goeden MA, Rose DJ, Mau B, Shao Y (1997) Science 277:1453
87. Pielak GJ (2005) Proc Natl Acad Sci USA 102:5901
88. Minton AP (2006) J Cell Sci 119:2863
89. Zhou HX, Rivas G, Minton AP (2008) Annu Rev Biophys 37:375
90. Minton AP (2001) J Biol Chem 276:10577
91. Dorsaz N, De Michele C, Piazza F, De Los Rios P, Foffi G (2010) Phys Rev Lett 105:120601
92. Zheng Y (2010) Nat Rev Mol Cell Biol 11:529
93. St Johnston D (2005) Nat Rev Mol Cell Biol 6:363
94. Birkenfeld J, Nalbant P, Yoon SH, Bokoch GM (2008) Trends Cell Biol 18:210
95. Good M, Tang G, Singleton J, Remenyi A, Lim WA (2009) Cell 136:1085
96. Seeliger MA, Kuriyan J (2009) Cell 136:994
97. Liu J, Nussinov R (2011) J Biol Chem 286:40934
98. Good MC, Zalatan JG, Lim WA (2011) Science 332:680
99. Csermely P, Palotai R, Nussinov R (2010) Trends Biochem Sci 35:539
100. Di Paolo G, De Camilli P (2006) Nature 443:651
101. Loose M, Kruse K, Schwille P (2011) Annu Rev Biophys 40:315
102. Elcock AH (2010) Curr Opin Struct Biol 20:196
103. Pielak GJ, Miklos AC (2010) Proc Natl Acad Sci USA 107:17457
104. Shaw AK, Pal SK (2007) J Photochem Photobiol B 86:199
105. Hu Z, Bowen D, Southerland WM, del Sol A, Pan Y, Nussinov R, Ma B (2007) PLoS Comput Biol 3:e117
106. Bhabha G, Lee J, Ekiert DC, Gam J, Wilson IA, Dyson HJ, Benkovic SJ, Wright PE (2011) Science 332:234
107. Agarwal PK, Billeter SR, Rajagopalan PT, Benkovic SJ, Hammes-Schiffer S (2002) Proc Natl Acad Sci USA 99:2794
108. Leventis R, Silvius JR (2010) Biophys J 99:2125
109. Nakano S, Karimata HT, Kitagawa Y, Sugimoto N (2009) J Am Chem Soc 131:16881
110. Xu Y, Wang H, Nussinov R, Ma B (2013) Protein charge and mass contribute to the spatio-temporal dynamics of protein-protein interactions in a minimal proteome. Proteomics. doi:10.1002/pmic.201100540. [Epub ahead of print]

Top Curr Chem (2013) 337: 139–164
DOI: 10.1007/128_2012_409

Published online: 7 March 2013

Allosteric Activation Transitions in Enzymes and Biomolecular Motors: Insights from Atomistic and Coarse-Grained Simulations

Michael D. Daily, Haibo Yu, George N. Phillips Jr, and Qiang Cui

Abstract The chemical step in enzymes is usually preceded by a kinetically distinct activation step that involves large-scale conformational transitions. In "simple" enzymes this step corresponds to the closure of the active site; in more complex enzymes, such as biomolecular motors, the activation step is more complex and may involve interactions with other biomolecules. These activation transitions are essential to the function of enzymes and perturbations in the scale and/or rate of these transitions are implicated in various serious human diseases; incorporating key flexibilities into engineered enzymes is also considered a major remaining challenge in rational enzyme design. Therefore it is important to understand the underlying mechanism of these transitions. This is a significant challenge to both experimental and computational studies because of the allosteric and multi-scale nature of such transitions. Using our recent studies of two enzyme systems, myosin and adenylate kinase (AK), we discuss how atomistic and coarse-grained simulations can be used to provide insights into the mechanism of activation transitions in realistic systems. Collectively, the results suggest that although many allosteric transitions can be viewed as domain displacements mediated by flexible hinges, there are additional complexities and various deviations. For example, although our studies do not find

M.D. Daily
Pacific Northwest National Laboratory, Richland, Washington 99352, USA

H. Yu
School of Chemistry, University of Wollongong, NSW 2522, Australia

G.N. Phillips Jr
Dept. of Biochemistry and Dept. of Computer Sciences, University of Wisconsin-Madison, WI 53706, USA

Dept. of Biochemistry & Cell Biology and Department of Chemistry, Rice University, Houston, TX 77006, USA

Q. Cui (✉)
Department of Chemistry and Theoretical Chemistry Institute, University of Wisconsin Madison, 1101 University Ave., Madison, WI 53706, USA
e-mail: cui@chem.wisc.edu

any evidence for "cracking" in AK, our results do underline the contribution of intra-domain properties (e.g., dihedral flexibility) to the rate of the transition. The study of mechanochemical coupling in myosin highlights that local changes important to chemistry require stabilization from more extensive structural changes; in this sense, more global structural transitions are needed to activate the chemistry in the active site. These discussions further emphasize the importance of better understanding factors that control the degree of co-operativity for allosteric transitions, again hinting at the intimate connection between protein stability and functional flexibility. Finally, a number of topics of considerable future interest are briefly discussed.

Keywords Allostery · Molecular motors · Enzyme catalysis · Molecular dynamics · Coarse-grained models · Small angle X-ray scattering · Co-operativity · Protein evolution

Contents

1 Introduction

For several decades, numerous experimental and computational studies have clearly illustrated that protein molecules exhibit motions that span a broad range of lengths and time scales. It is thus natural to ask what subset(s) of these motions are particularly important to the biological function of proteins. For enzymes, whose main biological function is to accelerate chemical transformations, the task becomes identifying motions that are intimately coupled to the chemical reaction. In addition to the fundamental importance, investigations along this line are also of practical significance since engineering necessary motions or flexibility into artificial enzymes is believed to be a key step to enhancing their catalytic proficiency to the level of natural enzymes. Despite progress made in the field of rational protein/enzyme engineering [1–3], most design studies rely on a framework that involves optimization of active sites in static structures. This is at least one important reason why computationally designed enzymes (e.g., abzymes [4]) are often substantially inferior in activity to naturally evolved enzymes. Therefore, although there is a growing awareness of the role of motions in catalysis, additional insight from integrated

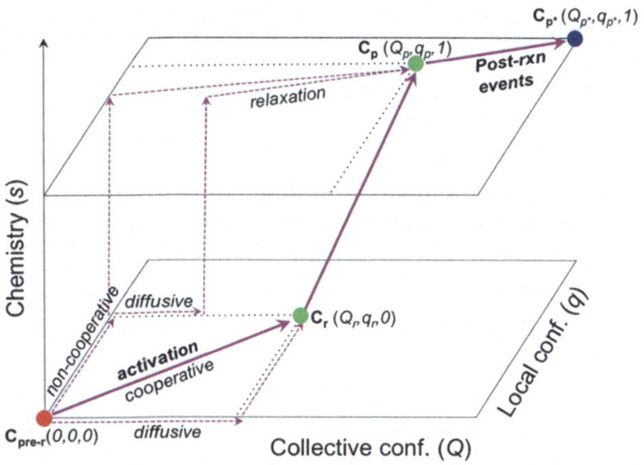

Fig. 1 A schematic sketch that illustrates the coupling between multi-scale structural changes (both collective and local conformational transitions) and chemistry in enzymes. The *filled circles* indicate kinetic states with different conformational (Q, q) and chemical (s) coordinates; the *thick arrows* indicate the dominant pathways that feature highly co-operative conformational transitions, while the *dashed arrows* indicate hypothetic pathways that are less co-operative and presumably have less flux. The "activation" process that goes from a pre-reactive conformation ($C_{\text{pre-r}}$) to a reactive conformation (C_{r}) may correspond to an open/close transition of an enzyme active site (e.g., in adenylate kinase) or the recovery stroke of myosin (see below); this activation process is the focus of this chapter. The chemical step may also involve structural transitions of multiple scales, as discussed in other studies [15]. Finally, the $C_p \rightarrow C_{p*}$ transition corresponds to structural changes following the chemical step in the catalytic cycle, such as opening of the active site for product release or the power stroke of myosin due to actin binding and release of inorganic phosphate

experimental and theoretic approaches is urgently needed before enzymes can be well designed de novo.

To help define the question more precisely, we note that, as illustrated schematically in Fig. 1, the chemical step in an enzyme is generally sandwiched between conformational change events that are kinetically distinct steps [5]. In a "simple" enzyme, the transition prior to the chemical step corresponds to the closure of the active site upon substrate binding, and the conformational transition following the chemical step corresponds to opening of the active site for product release; in more complex enzymes, such as molecular motors (see below), these transitions before and after the chemical step are more complex in nature and coupled to the binding/association of other protein partners, such as actin or microtubule. All three steps, including the chemical step itself, may involve structural transitions that span multiple scales, as Fig. 1 attempts to highlight using two abstract conformational coordinates, Q and q. Therefore, by asking "what motions are intimately coupled to chemistry in an enzyme," one may study any one of these three steps [6]; in this context, it is worth recalling that the rate-limiting event for a catalytic cycle does not have to be the chemical transformation itself.

Since efficient chemical reactivities are the defining feature of enzymes, it's not surprising that much attention has been given to the elucidation of motions that are directly involved in the chemical step; in part, this is because most enzyme catalyzed chemical reactions are fairly local in nature and thus the idea that they are coupled to much larger scale motions is intriguing. Along this line, both experimental and theoretical studies have made great progress in the past decade, especially in the context of proton/hydride transfer reactions (collectively referred to as H-transfer below) in several enzyme systems [7–10]. As reviewed by other chapters in this special issue, the motions tightly coupled to such H-transfer reactions are mostly rather fast and localized vibrations (e.g., at frequency $\sim$200 cm^{-1}) that modulate the barrier crossing transmission coefficient [11–13]. By comparing the conformational ensembles for the reactant state and transition state from potential of mean force (PMF) simulations, more collective motions have been suggested to determine the free energy barrier and therefore be "conducive" to H-transfer reactions [14, 15]; the causality relation between such motions and the H-transfer process, however, is not always straightforward to determine.

In our recent studies we have been focusing on slow motions in enzymes that we term "activation" transitions in Fig. 1. It is important to study these motions because they are required to bring catalytic motifs to proximity so that the chemical step can occur efficiently; since multiple structural rearrangements are involved, one question of interest here is what rearrangements are in fact most important to the activation of the subsequent chemical step, and another general question is what factor(s) control the couplings among the various structural rearrangements and the overall rate of the transition. Unlike the motions that occur during the chemical transformation, as discussed in the study of H-transfer reactions, the activation processes of interest correspond to transitions between well-defined kinetic states of the enzyme–substrate complex (e.g., $C_{\text{pre-r}} \rightarrow C_{\text{r}}$ in Fig. 1) and therefore are mostly in the microsecond to millisecond time scale. They can be studied directly with various experimental approaches, which provide important information for and the opportunity to test theoretical analyses.

To study specific examples, we have chosen biomolecular motors and signaling proteins in our studies. Our choice is motivated by the consideration that a tight mechanochemical coupling (i.e., coordination between chemistry and conformational transitions) in these systems is likely essential to their biological function (i.e., high efficiency for energy/signal transduction [16, 17]); thus it is particular worthwhile to understand the mechanism through which the chemical step (e.g., phosphorylation or ATP hydrolysis) is coupled with the multitudes of conformational rearrangements. From a biomedical perspective, such research is also significant because many mutations that affect the mechanochemical coupling are involved in serious diseases. For example, mutations that perturb the coupling between the ATPase activity and the recovery stroke in the motor myosin (see below) are known to cause hypertrophic or dilated cardiomyopathies [18]. Similarly, mutations that modify the response of kinases to phosphorylation are implicated in various cancers [19, 20].

Regarding another model system, we focus on the open/close transition in the enzyme adenylate kinase (AK). Kinetic analysis has established that the open/close transition is rate-limiting for several bacterial AKs [21], and the relatively small size of AK makes it an ideal system for in-depth analyses regarding factors that control the rate of large-scale transitions in enzymes [22–24]. Indeed, the system has been studied by many computational approaches at both atomistic [25–31] and coarse-grained (CG) levels [32–38]. Nevertheless, as we discuss below, the discrepancies that remain between different models highlight several fundamental issues regarding the mechanism of large-scale transitions in proteins.

In the following, we first briefly discuss the computational methods used in our studies of slow activation transitions in enzymes; for another complementary review that focuses on other computational techniques for studying functional transitions in small signaling proteins and ion channels, see [39]. Next, we discuss a few key results concerning the mechanochemical coupling in myosin and the open/close transition in AK from our recent work [40–46]; although normal mode analysis [47] was found useful in our analysis of myosin [48], we will not review the approach/results here since the topic will be the focus of other authors in this issue. Finally, we summarize the key conclusions and also comment on future directions of research.

2 Computational Methods

Two key objectives for a computational study of functional transitions are (1) to identify the factors that dictate the spatial scale and rate of the transition and (2) to identify the functional consequence of different components of the underlying motions. For the second objective, computational studies are particularly useful because one can construct in silico models that include only a subset of motions (which is difficult to accomplish with experiments) and then explicitly evaluate the functional consequence, such as by doing QM/MM calculations for the subsequent chemical step; this is illustrated below with the example of ATPase activation in the molecular motor myosin. To achieve the first objective it is important to characterize the transition pathway(s) and, arguably more importantly, the transition state ensemble for the transition of interest. Since the underlying motions of interest are in the microsecond to millisecond regime, this is a challenge difficult to meet with straightforward atomistic simulations, although advances are being made in both computational hardware [49, 50] and/or sampling techniques (also see Sect. 4). With the standard computational facilities, the most practical strategies for large proteins remain atomistic simulations with carefully chosen biases and coarse-grained simulations.

In biased atomistic simulations, an additional biasing potential is used to drive the relevant conformational transition to occur during the time scale accessible to computations. The bias can be applied either as a restraint or a holomonic constraint, leading to biased (BMD [51]) or targeted molecular dynamics (TMD [52, 53]), respectively; BMD is also similar to the steered molecular dynamics (SMD [54]). For example, in TMD the RMSD of the structure relative to a target conformation is constrained to decrease in a monotonic fashion as a function of time; by design, the structure evolves toward the target in a defined time scale. Compared to the alternative approach of minimum energy path (MEP) [55] or related method (e.g., maximum flux path [56]), the advantage of TMD/BMD/SMD is that thermal fluctuations and therefore entropic factors are (partially) taken into consideration. On the other hand, due to the presence of biases, the results of these simulations have to be interpreted with care, especially when the time scale of the simulation is much faster than the realistic time scale. An additional factor to consider is the coordinate being biased, which is often chosen based on intuition, such as relative RMSD ($\Delta RMSD$), rotational angles between domains. The choice of the bias coordinate may have a non-trivial impact on the observed sequence of motions [57]. For example, a collective coordinate such as $\Delta RMSD$ is likely to encourage large-scale motions that greatly reduce the $RMSD$ values over more local motions. Therefore, the causality of different motions from biased MD simulations should be interpreted with great care. At this point, these biased MD simulations are best thought of as approximate means to identify transient interactions not readily detected from structures that correspond to the end states of the transition, and the kinetic relevance of these transient interactions is best tested with experimental studies. Better understanding the causality of different motions requires computing the underlying free energy landscape, or the PMF [58] along the relevant coordinates using umbrella sampling [59] or other sampling techniques [60, 61], although this is only possible when the number of active degrees of freedom is relatively small [62, 63]. Another alternative is to sample directly the transition pathway(s) in an unbiased fashion using advanced techniques such as milestoning [64], thermal string methods [65], and multi-state Markov models [66, 67] (see Sect. 4).

The complementary approach is to use coarse-grained (CG) models [68], which are computationally efficient and therefore can be used to sample the relevant slow motions without additional biases. Compared to CG models for lipids and DNA, however, generally reliable and transferrable CG models for proteins are not yet available [69]. Thus most CG studies of protein conformational transitions employ structure based (or native-centric) models, also referred to as Go models. Different strategies have been proposed and applied to construct such models for more than one conformational state [70, 71]. The underlying assumption behind these structure-based CG models is that protein motions are largely dictated by the structural topology of the system rather than detailed energetics [72]. Despite the apparent success of such models in application to several protein and protein-RNA/DNA systems [68, 73], establishing relevant experimental verifications of these models and their predictions remains an active area of research.

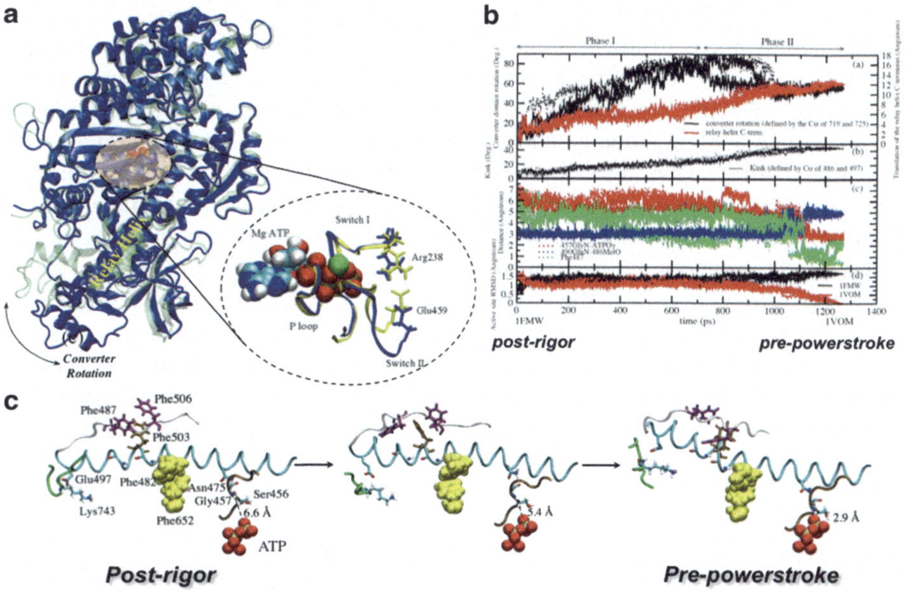

Fig. 2 The recovery stroke of myosin II, which is the conversion between the post-rigor and pre-powerstroke kinetic states. (**a**) Structural differences between two X-ray conformations (post-rigor [81], in *blue*, and pre-powerstroke [82], in *green*) of the *Dictyostelium discoideum* myosin motor domain. The most visible transition is the converter rotation, although there are also notable changes in the nucleotide binding site and the relay helix that connects the converter and the nucleotide binding site [77]. With ADP $\cdot$ VO_4^- bound, the nucleotide binding site of the pre-powerstroke state has a closed configuration (in *yellow*); with ATP bound, the nucleotide binding site in the post-rigor state is open (in *blue*) due to displacement of the Switch II loop. (**b**) Variation of critical structural parameters along the three TMD trajectories for illustrating the sequence of events along the approximation transition paths for the recovery stroke [41]. (**c**) Three snapshots (at 0.0 ps, 630.0 ps, 1270.0 ps) from one of the three TMD simulations, in the same format as Fig. 2 in [78] for comparison, to illustrate the proposed coupling between the small motion of SwII with the large translation of the relay helix C-terminus

3 Results and Discussion

In this section we briefly review two systems that we have studied recently: myosin and adenylate kinase; they illustrate the value of atomistic and CG simulations in the analysis of slow functional transitions (activation) in proteins, respectively.

3.1 Myosin

Myosin is a family of molecular motors that play various essential roles such as cellular transportation and cell division [74]. The one that we have analyzed is

Myosin II, simply referred to as myosin below, which is involved in muscle contraction. It is one of the best characterized motors at the kinetic and structural levels [75–77]. The two kinetic/conformational states of relevance here are the "post-rigor" and "pre-powerstroke" states of the motor domain, which differ in both the nucleotide binding domain and the converter domain as well as the intervening structural motifs, such as the relay helix (Fig. 2a); the two domains are separated by more than 40 Å and rotation of the converter is further propagated into the striking displacement of the lever arm that is most visible in single molecule studies of processive myosin motors. The transition between the two states, referred to as the "recovery stroke," occurs on the 10-ms scale, and the hydrolysis of ATP is believed to occur only in the pre-powerstroke conformation. A thorough understanding of mechanochemical coupling in myosin, therefore, requires elucidating the detailed transition pathway between the "post-rigor" and "pre-powerstroke" states and establishing what subset(s) of motions during this recovery stroke are most important to the activation of the ATPase activity.

3.1.1 Approximate Pathways for the Recovery Stroke

To probe the mechanism of the recovery stroke, several computational studies have been carried out. Fischer and co-workers [78] have calculated a minimum energy path (MEP) that connects the two conformational states, and the results pointed to a two-phase transition mechanism that initiates from a hydrogen-bond formation near the active site (between Gly457 and the γ-phosphate of ATP) and propagates sequentially through the relay helix to the converter domain. A set of hydrophobic interactions that form during the transition were proposed to stabilize the local unwinding of the relay helix, which ultimately leads to the translation/rotation of the C-terminal helix and converter. A recent set of milestoning calculations [79] based largely on the MEP pathway as the initial guess led to an estimated time scale for the transition that is consistent with the experimental transition rate.

Using a rather different TMD approach and an implicit solvent model, we found that most rotation of the converter domain occurs in the first stage of the transition while structural changes in the relay helix and complete closure of the active site occurs at a later stage to stabilize the converter conformation via a series of hydrogen bonding interactions as well as hydrophobic contacts (Fig. 2b) [41]. For example, the "unwinding" of the relay helix happens much later in the TMD simulations compared to the MEP description. In contrast to the MEP results, which suggest that the kink and unwinding in the relay helix are induced by Switch II (SwII) closure via a single hydrogen-bonding interaction from Asn 475, the TMD simulations tend to suggest that the converter rotation, via strong polar interactions to the relay helix and the relay loop, induces the formation of the hydrophobic cluster halfway in the relay helix as well as some polar interactions (e.g., Asn483-Glu683) to produce the kink in the relay helix. The interactions between the Sw II and the relay helix stabilize, in return, the new conformation of the relay helix.

Since there are major approximations in all reported computational studies so far, it remains unclear which mechanism dominates in reality; the MEP study used a simpler potential function and does not include thermal fluctuations of the protein, while the TMD approach uses an approximate biasing coordinate ($\Delta RMSD$) that may encourage collective motions (e.g., converter rotation) in the early stage of the transition. The encouraging aspect is that both TMD and MEP studies point to the importance of a consistent set of hydrophobic interactions (e.g., between Phe482, Phe487, Phe503, Phe506, and Phe652) and hydrogen bonding/salt bridge interactions (e.g., Glu497-Lys743) between the relay helix, active site, and the converter domain (see Fig. 2c); it is worth noting that mutations involving several of these corresponding residues in human cardiac myosin (e.g., Phe506, Glu497) are known to cause cardiac contractile dysfunctions. These discussions highlight both the value of these approximate computational techniques and the need for developing novel methods that allow a quantitative computational analysis of slow motions in biomolecules.

We have also carried out PMF calculations [43] for the recovery stroke process with ATP as the ligand using the $\Delta RMSD$ reaction coordinate defined using the crystal structures of the "post-rigor" and "pre-powerstroke" states. Analysis of key geometrical properties in different windows found very similar trends as in the TMD simulations, indicating no major mechanistic change between the non-equilibrium TMD simulations and the umbrella sampling simulations that are closer to equilibrium (in total $\sim$50 ns with an implicit solvent model). Overall, the calculated PMF is largely downhill in nature and reveals a rather broad basin around the pre-powerstroke state; this is qualitatively consistent with results from normal mode analysis as well as the experimental observation that the pre-powerstroke state has a significant degree of flexibility in the lever arm/converter. The downhill nature is qualitatively similar to the PMF results of a study using myosin II from a different organism and explicit solvent simulations [80], although the degree of exothermicity is significantly smaller in our result. The quantitative nature of these PMF results is unclear given the scale of the structural transition and various approximations inherent in this type of analysis. Nevertheless, the results suggest that the recovery stroke is largely diffusive in nature and doesn't involve any major energetic bottleneck; this is qualitatively consistent with the observation from TMD/MEP simulations that multiple polar and hydrophobic interactions break and form continuously throughout the recovery stroke.

3.1.2 Coupling Between Conformational Changes and ATP Hydrolysis

To evaluate the functional impact of motions implicated in the recovery stroke, we have carried out QM/MM simulations for ATP hydrolysis using snapshots collected from MD simulations of not only the two relevant X-ray structures [81, 82] but also a hybrid conformational state in which the active site in the post-rigor X-ray structure was closed in silico by displacing the commonly discussed SwII loop [76, 77]. This was meaningful to do because our PMF calculations [40] showed that SwII closure has a rather flat free energy profile in the post-rigor state. The goal is to establish explicitly whether only structural changes in the immediate neighborhood

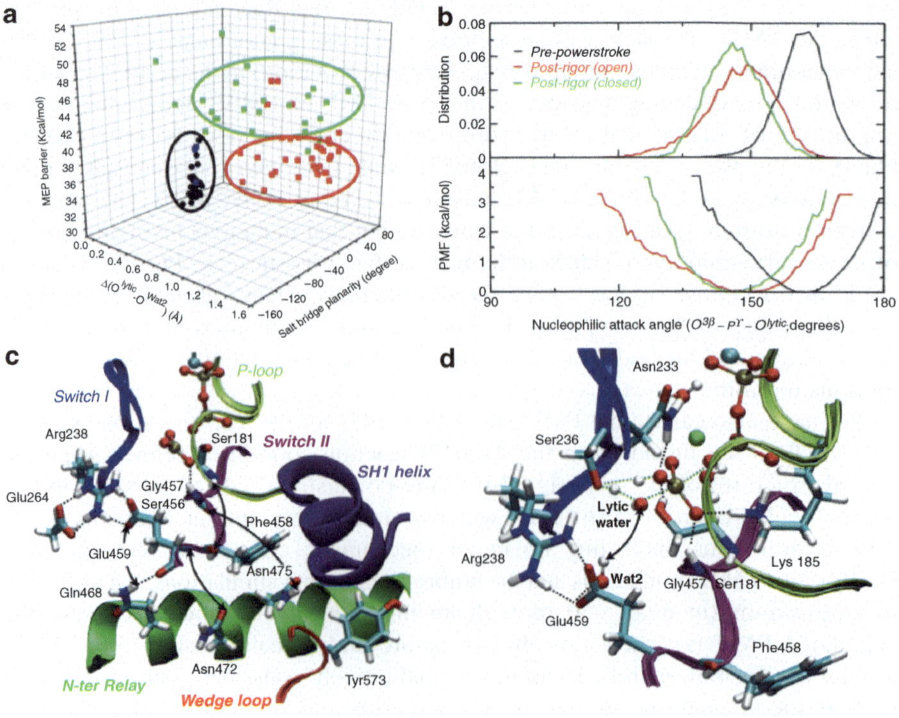

Fig. 3 The dependence of ATPase activity of myosin on the structural state of the motor domain, i.e., the mechanochemical coupling in myosin [40, 42]. (**a**) Minimum energy path (MEP) barriers for the first step of ATP hydrolysis calculated starting from snapshots collected from equilibrium simulations of the pre-powerstroke state and a closed post-rigor structure (see text). The barriers are plotted against the Arg238-Glu459 salt bridge planarity and the differential distance between the lytic water and Wat2 (see **d**) in the reactant and transition state. The *black dots* indicate data for the pre-powerstroke state; the *blue*, *green*, and *red* sets indicate data from the closed post-rigor simulations with different behaviors of Wat2 [42]. Note that the MEP barriers are systematically higher than the free energy barrier due to the lack of sampling specific local rearrangements; see [42] for discussions. (**b**) Comparison of the distribution and corresponding PMF for the nucleophilic attack angle based on equilibrium simulation for the pre-powerstroke state and two post-rigor structures. (**c**) Key hydrogen-bonding interactions in the active site region of the closed post-rigor structure. The *arrows* indicate interactions that are broken when SwII is displaced to close the active site in the post-rigor structure; i.e., rearrangements in the N-terminus of the relay helix and wedge loop are required to form the stable active site as in the pre-powerstroke state. (**d**) A representative active site structure for the ATP hydrolysis transition state with a twisted Arg238-Glu459 salt-bridge configuration. The notable feature is that Wat2 in the active site remains hydrogen bonded to Glu459 and therefore does not provide the critical stabilization for the transition state

of ATP are sufficient to activate the hydrolysis activity. This is only possible to do computationally and clearly illustrates the unique value of computational studies in establishing the functional relevance of specific motions.

Remarkably, QM/MM calculations [42] found that ATP hydrolysis tends to have very high barriers even though the key residues directly in contact with the γ phosphate have the same average configuration as in the pre-powerstroke state (Fig. 3a). Part of this is due to the difference in the key "nucleophilic attack angle" ($O^{3\beta}$ -P^{γ} -O^{lytic}), the distribution of which peaks around 165° in the pre-powerstroke state but around 150° in both the post-rigor and the closed post-rigor conformations. Since the average nucleophilic attack angle in the transition state is ~169°, the free energy penalty associated with properly aligning the lytic water in both the post-rigor and the closed post-rigor conformations is on the order of 2–3 kcal/mol (Fig. 3b). Although this is not a small contribution in kinetic terms (corresponds to 30- to 150-fold change in the rate constant at 300 K according to transition state theory), it is clear that the nucleophilic attack angle is not as dominating a factor as commonly suggested for dictating the hydrolysis activity [83, 84].

Further analysis found that the ATP hydrolysis barrier in the closed post-rigor conformation is higher because displacing SwII alone to close the active site in the post-rigor state leaves many crucial interactions unformed or even breaks existing interactions (Fig. 3c). For example, although Gly457 forms a stable hydrogen bonding interaction with the γ phosphate of ATP upon SwII displacement, the interaction between Ser456 main chain and Asn475 in the relay helix is broken; the interaction between Gly457 and γ phosphate also becomes weaker in the later segment of the simulations. Similarly, although Glu459 forms a salt-bridge interaction with Arg238 when SwII is displaced, the interactions between the main chain of Glu459 and Asn472 in the relay helix are lost; instead, the carbonyl of Glu459 forms a hydrogen bond with the sidechain of Gln468. As a result, the extensive hydrogen-bonding network that involves Arg238, Glu459, Glu264, and Gln468 observed in the pre-powerstroke state is not present in the closed post-rigor state, which explains the higher rotational flexibility of Glu459 in the latter conformational state. Finally, since Tyr573 in the "wedge loop" [85] remains far from SwII in the closed post-rigor state, there is ample space for Phe458 to sample multiple rotameric states and its main chain interaction with Ser181 remains unformed, in contrast to the situation in the pre-powerstroke state. Although these additional structural flexibilities seem fairly subtle, analysis indicates that they have a significant impact on the hydrolysis barrier. For example, as Glu459 rotates out of the salt-bridge plane, the second active-site water (Wat2) tightly associated with its sidechain forms a hydrogen-bonding interaction with the lytic water only in the reactant state but not the transition state of hydrolysis (Fig. 3d). Accordingly, Wat2 makes an unfavorable contribution (~6 kcal/mol) to the hydrolysis barrier.

Therefore the emerging picture is that the transition from the post-rigor state to a structurally stable closed active site (which apparently is critical to efficient ATP hydrolysis) relies on not only the displacement of SwII but also more extensive structural rearrangements in the nearby region. Without the latter, residues in the "second coordination shell" of the γ phosphate may adopt configurations that hamper the effective hydrolysis of ATP. In other words, structural transitions remote from the active site can play an active role in regulating the hydrolysis of ATP, rather than passively responding to structural changes in the active site; this is

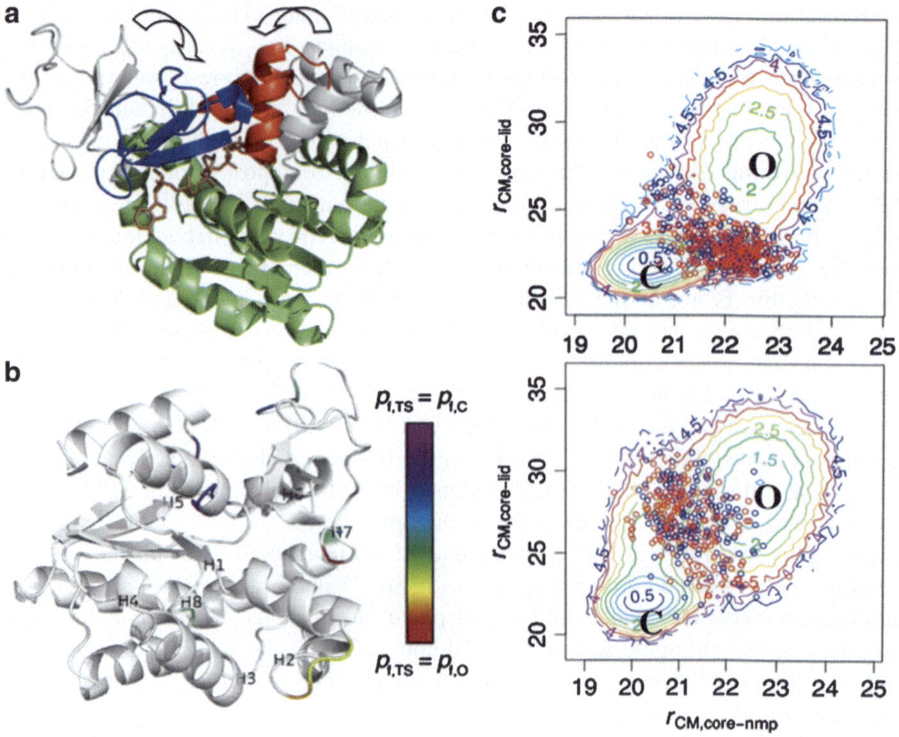

Fig. 4 Coarse-grained (CG) simulation results [44] for the open/close transition of *E. coli* adenylate kinase (AK). (**a**) Open (*gray*) and closed (*colored*) conformations are superimposed by the coordinates of the CORE domain (*green*). *Blue*: LID domain, *red*: NMP-binding domain, *brown*: bisubstrate analog AP_5A. (**b**) Fractional progress of folding in the transition state (TS) ensemble relative to the open and closed state ensembles. For residues that differ substantially in flexibility (characterized by p_{fold} in [44]), the flexibility in the TS ensemble is contoured on a rainbow scale from *red* (0) to *purple* (1); the results highlight that different regions contribute differently to the entropic barrier of the transition. Labels H1–H8 indicate the respective central residues of the eight hinges identified in [23]. (**c**) Potentials of mean force that contrast the free energy landscape for NMP vs LID motions in ligated and apo simulations. The *circles* indicate the location of representative TS ensembles collected during the simulations

likely a general feature shared among biomolecules whose function relies on a tight mechanochemical coupling between distant sites.

3.2 Adenylate Kinase

In recent years, AK has emerged as a prototypical system and its chemically rate-limiting open/close transition (Fig. 4a) has been subjected to numerous experimental [21–24, 86, 87] and computational [25–38] analyses; still, factors that dictate the open/close transition rate remain elusive. For example, dynamic importance sampling [31] has suggested that a few contacts in the LID-NMP interface region progressively "zip" in the closing transition and thus determine the open/close rate. In addition, normal-mode analysis [36] and a mixed Go model [33, 34] have suggested that a few residues must locally unfold or "crack" to relieve strain associated with the LID and NMP domain motions; the two studies, however, pointed to different stressed regions. By contrast, recent experiments have suggested a different (but not exclusive) mechanism wherein functionally important dynamics are distributed among many residues. For example, swapping the entire LID and NMP domain sequences (but not just the CORE-LID hinges) between homologous mesophilic and thermophilic AKs interconverts catalytic properties (and, thus, presumably the open/close rate) [24].

Since the time scale of the transition is milliseconds and the transition process is complex, atomistic simulations have been limited to the elucidation of the most salient features of the transition (e.g., intrinsic flexibility of the LID domain, local flexibility of hinges, and the role of ligands in stabilizing the closed conformation) and coarse-grained models have been found invaluable for probing the transition mechanism. In the following, we summarize results from our CG studies [44, 45] and also initial efforts to validate the CG model using small angle X-ray scattering (SAXS) data [46].

3.2.1 CG Models for the Open/Close Transitions

We have carried out detailed analysis of the transition state (TS) ensemble for the open/close transition using a double-well Go model with and without pseudo-contacts added to the closed potential to simulate ligand binding; the approximate TS ensemble was collected using the criterion of Hummer and Best [88]. By simultaneously characterizing the contributions of rigid-body (Cartesian), backbone dihedral, and contact breaking/formation motions to the TS structure and energetics, we were able to predict specific residues and contacts that influence the open/close transition rate. For example, we found that backbone fluctuations are reduced in the open/close transition in parts of all three domains. Among these "quenching" residues, most in the CORE domain, especially residues 11–13, are rigidified in the TS of the ligated simulation and thus slow down the open/close transition by entropically raising the free energy of the TS relative to the native states, while residues 42–44 in the NMP domain are flexible in the TS and thus facilitate the open/close transition (Fig. 4b). In contact space, in both unligated and ligated simulations, one nucleus of closed-state contacts includes parts of the NMP and CORE domains. These results allowed us to predict mutations that will perturb the opening and/or closing transition rates by changing the entropy of dihedrals and/or the enthalpy of contacts [44]. Considering the approximate nature of the CG model, we note that the use of "enthalpy" and "entropy" factors here is qualitative and largely reflect whether inter-residue interactions or changes in thermal fluctuations make the dominant contribution.

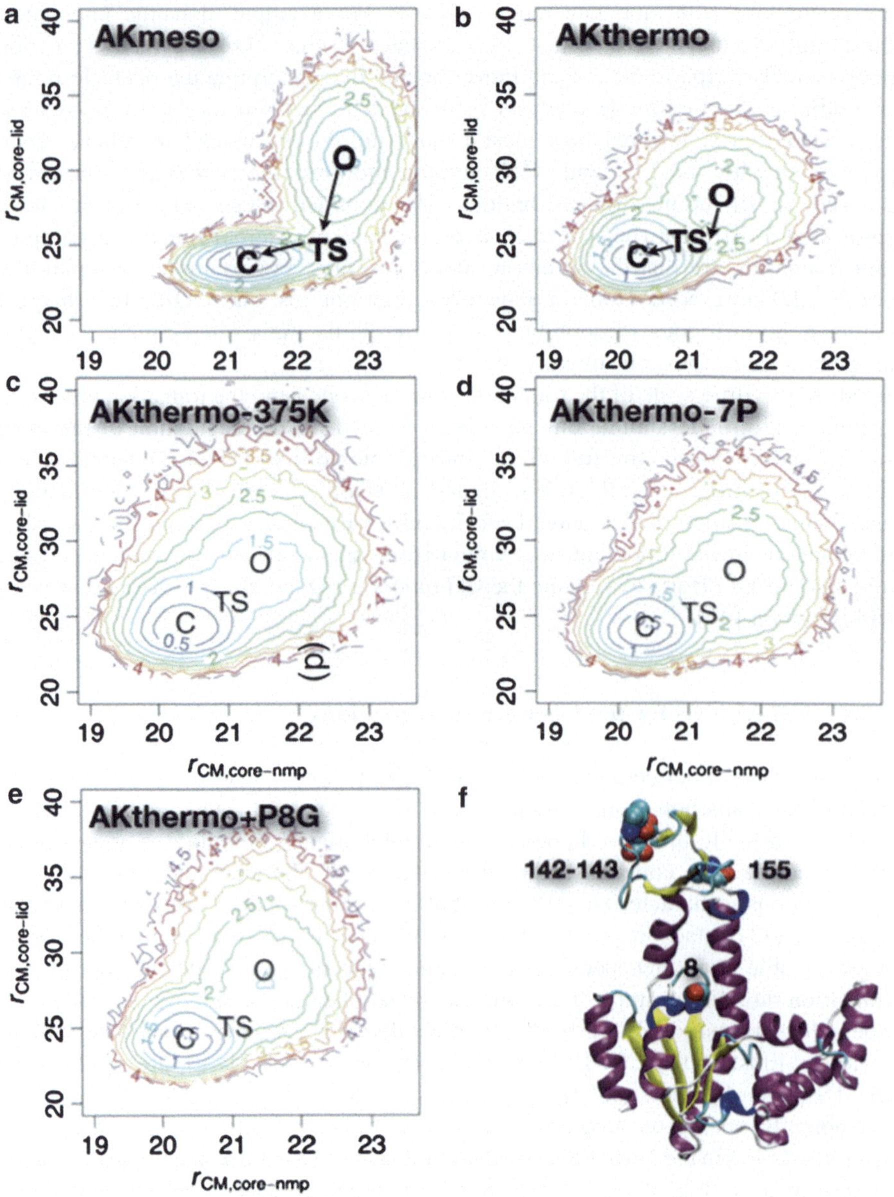

Fig. 5 Potentials of mean force for the open/close transition in mesophilic AK (*E. coli*, AKmeso), thermophilic AK (*Aquifex aeolicus*, AKthermo), and various in silico mutants of AKthermo using the double-well Go CG model [45]. AKthermo-375 K indicates that the simulation is carried out at 375 K instead of 300 K (all other panels); AKthermo-7P indicates the AKthermo mutant in which all Pro residues unique to AKthermo are changed to the corresponding ones in AKmeso; AKthermo+P8G is the AKthermo mutant in which only Pro 8 (see position in the structure in (**f**) is changed to a Gly. These results illustrate that AKmeso/AKthermo approximately satisfy the "corresponding state hypothesis" [90, 91], and that Pro 8 is particularly important in controlling the flexibility of the LID domain in the open state

Moreover, we observed that LID closure precedes NMP closure in the ligated simulation, consistent with other coarse-grained models of the AK transition [34]. However, NMP-first closure is preferred in the unligated simulation (Fig. 4c), which highlights that ligand binding can not only stabilize the closed conformation but also alter the kinetic pathway of closure. In this case, the driving factor for the pathway switch is likely enthalpic, since ligand binding adds many more contacts to the CORE-LID interface than to the CORE-NMP interface in the TS.

Another interesting observation from our study [44] is that "cracking" [33, 34] was not necessarily involved in the open/close transition of AK. In the TS ensemble of our model, for all dihedral degrees of freedom, folding probability lies approximately within the interval set by the open and closed states; i.e., no local unfolding unique to TS is implicated. This difference in results from previous CG studies [33, 34] probably arises from differences between the dihedral potentials in our work and that of Whitford and co-workers [34]; although the dihedral potential of Whitford et al. is specific to the open state, the dihedral potential in our model (that of Brooks and co-workers [89]) is sequence specific but generic to the native structures. Since our CG model is also approximate in nature, our observations do not necessarily rule out the relevance of cracking in large structural transitions; our results suggest that such transitions do not have to invoke local unfolding and highlight the importance of carefully validating the CG model for the problem of interest (see below).

In a more recent study [45], we have applied a similar computational framework to analyze comparatively the structural transitions in the mesophilic (*Escherichia coli*) and thermophilic (*Aquifex aeolicus*) AK enzymes (AKmeso, AKthermo). Experimentally, the latter was shown to have a lower opening rate than AKmeso at room temperature; the closing rates were more similar [22, 86]. Computationally [45], the double-well Go model found that AKmeso and AKthermo share a LID-first closure pathway in the presence of ligand, although LID rigid-body flexibility is considerably less in the open state ensemble of AKthermo than in that of AKmeso (Fig. 5a, b). Backbone foldedness in the open state and/or transition state ensembles increases significantly relative to AKmeso in some interdomain backbone hinges and within LID. In contact space, the TS of AKthermo has fewer contacts at the CORE-LID interface but a stronger contact network surrounding the CORE-NMP interface than the TS of AKmeso. Consistent with the corresponding states hypothesis [90, 91], increasing the simulation temperature of AKthermo increases LID rigid-body flexibility in the open state ensemble (Fig. 5c).

We have also attempted to use the CG framework to probe whether we can computationally interconvert the motional characteristics of AKthermo and AKmeso. Motivated by the discussions of Henzler-Wildman et al. [23], we have also focused on the Pro residues unique to AKthermo. Although computational mutation of seven prolines in AKthermo to their AKmeso counterparts produced the expected perturbations (Fig. 5d), mutation of these sites, especially positions 8 and 155 (see Fig. 5f), to glycine was required to achieve LID rigid-body flexibility and hinge flexibilities comparable to AKmeso (Fig. 5e). Analysis of the impact of these mutations on rigid body motion, dihedral flexibility, and inter domain contacts suggested that the contacts between CORE and CORE-LID connector helix 1 are likely the most important for modulating the global transition. Interestingly, other mutants spatially close to the P8 sites have been shown in recent experiments to have significant functional effects. For example, glycine mutants that destabilize the CORE-LID connector helices (I116G + L168G) increase ATP binding affinity by increasing the open to closed equilibrium constant [87]. On the other hand, mutating the seven sites to proline in AKmeso reduces some hinges' flexibilities, especially hinge 2, but does not reduce LID rigid-body flexibility, suggesting that these two types of motion are decoupled in AKmeso. Therefore, our results suggest that hinge flexibility and global functional motions alike are correlated with but not exclusively determined by the hinge residues.

3.2.2 Validation of the CG Model with SAXS

Although the Go-based CG simulations are informative and can stimulate new experiments, it remains a fundamental challenge to validate a CG model for complex biomolecules. For example, parameters in our model were calibrated to make RMSD from the CG simulations of individual open and closed states fit atomistic MD results [25]. It is possible, however, that these ~50-ns atomistic simulations underestimate the true flexibility of the system at the biologically relevant μs-ms time scale, as hinted by H/D exchange data from a recent NMR study [92]. Establishing proper benchmarks for CG models and improving their robustness remains an active and fruitful area for theoretical research.

As a useful step in this direction, we have recently carried out a study [46] that combines CG simulations and SAXS, taking advantage of a recently developed algorithm for computing SAXS profiles using residue-level CG models for proteins [93]. We had several aims. First, to estimate global flexibility for AK, we compared experimental SAXS curves to those calculated from CG simulation ensembles of AK using different strengths of inter-residue interactions. Second, to identify possible population shifts, we fitted SAXS curves measured in both the absence and the presence of ligand to linear combinations of predicted scattering from the open (O) and closed (C) state simulations (i.e., $I_{\mathrm{comb}} = w_{\mathrm{O}}\, I_{\mathrm{O}} + w_{\mathrm{C}}\, I_{\mathrm{C}}$). Specifically, we fitted log[I_{comb}] to experiment using linear regression over the range $0.14 < q < 0.3$. Finally, as discussed in [46] but not here, we have also calculated the correlation between predicted scattering and various structural metrics over large simulation ensembles, suggesting that scattering is most sensitive to the CORE-LID distance; this provides a way to interpret SAXS data better at the structural level.

As discussed in our previous work [44], we calibrated the simulated flexibility of AK by varying the contact energy scale (S_{con}), by which we scale the Karanicolas/Brooks energies [89] to compensate for extra backbone bond angle potential. For our open/close conformational transition simulations, we calibrated S_{con} to 2.5 so that the closed state simulation averaged about 2.0 Å Cα RMSD with respect to the closed crystal structure to reproduce prior atomistic simulations of AK. This

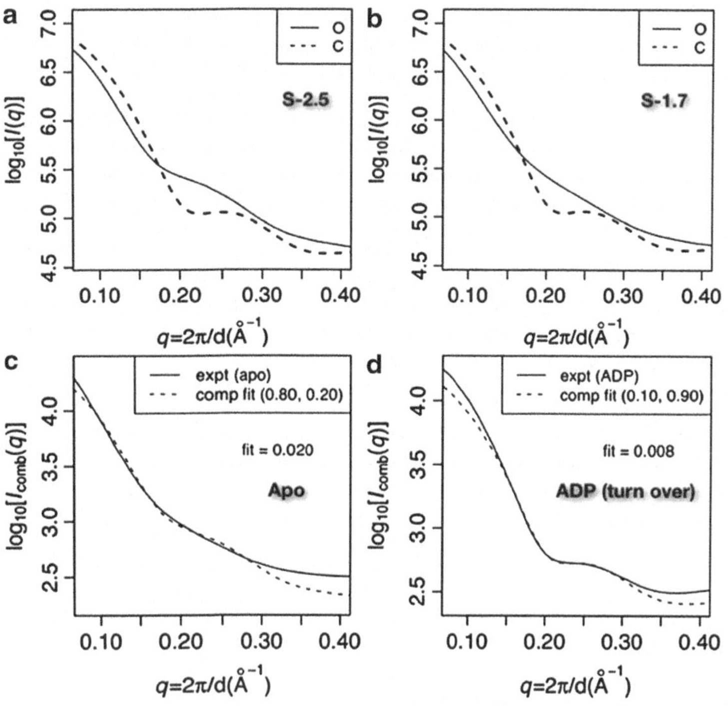

Fig. 6 Experimental vs predicted Small Angle X-Ray Scattering (SAXS) profiles for various states of AK [46]. (**a**, **b**) Predicted scattering at contact energy scales (S_{con}) of 2.5 and 1.9, respectively. The solid (*dashed*) line indicates the prediction from the open (close) simulation. For each calculation, predicted scattering ($I(q)$) is averaged over 1,000 randomly selected structures from the corresponding simulation ensemble, and $\log_{10}[I_{avg}(q)]$ is plotted. (**c**) Fits of $\log_{10}(I_{comb})$ at $S_{con} = 2.5$ to the *apo* experimental data over $0.14 < q < 0.3$. I_{comb} is the optimal linear combination of predicted scattering for the open (O) and closed (C) states to fit the data; the weights (w_O, w_C) are indicated in parentheses. The "fit" in the panel indicates the root mean square deviation between experimental and fitted computational data over that q range. (**d**) The fit of $log_{10}(I_{comb})$ at $S_{con} = 1.9$ to the ADP-bound (turning over condition) experimental data. Note that the experimental data were collected for *B. globisporus* AK while all calculated ensembles were collected using the *E. coli* AK; SAXS patterns from *E. coli* and *B. globisporus* AKs correspond closely under conditions where data from both species are available (data not shown)

significantly exceeded the S_{con} of 1.7 that was used in double-well Go simulations of smaller conformational transitions [71, 94].

Figure 6a shows that with $S_{con} = 2.5$, the closed state simulation predicts a more curved SAXS profile than the open state simulation, especially near $q \sim 0.22$ Å^{-1}; this is consistent with the closed state being more ordered than the open state as expected. With $S_{con} = 1.9$ (Fig. 6b) the predicted open state scattering profile is substantially less inflected near $q \sim 0.22$ Å^{-1} than at $S_{con} = 2.5$. Conversely, the predicted closed state curves are similar for ensembles generated using the two S_{con} values, both exhibiting a small dip near $q \sim 0.22$ Å^{-1}.

Figure 6c shows that the optimal fit over $0.14 < q < 0.3$ Å^{-1} occurs for $S_{con} = 1.9$ with $w_O = 90\%$ for the apo condition; with $S_{con} = 2.5$, the predicted scattering is more strongly inflected at $q \sim 0.22$ Å^{-1} than the experimental data, while at $S_{con} = 1.5$, the predicted scattering curve is substantially shallower. This provides a measure of the impact of the strength of inter-residue interactions on the intrinsic flexibility of the system and the corresponding sharpness of features in the scattering pattern. We also note that the $S_{con} = 2.5$ open state simulation produces a slightly better fit at low angles ($q < 0.14$) than the $S_{con} = 1.9$ open state simulation. The relatively dominant population of the open state in the absence of any substrate makes biochemical sense, although it is worth noting that the single molecule FRET study suggested that the closed state is dominant even in the absence of substrates.

Figure 6d shows the fit of I_{comb} at $S_{con} = 1.9$ to experimental data collected in the presence of substrate ADP (i.e., turning over condition). A 10% population of the open state flattens the small dip predicted at $q \sim 0.22$ Å^{-1} (Fig. 6b), producing a very small RMSD for the fit. The significant population shift going from the apo to the ADP condition is biochemically consistent with the conformational equilibrium required for catalytic cycling. By contrast to the apo experimental data, a broad range of S_{con}, especially between 1.7 and 2.1, produces good fit to the ADP data.

4 Concluding Discussions and Future Outlooks

Enzyme catalysis is a multi-step process that involves complex interplay of chemistry and conformational motions that span many temporal and spatial scales. To identify motions that dictate the catalytic function of enzymes, it is imperative to define clearly which step of the catalytic cycle is of interest (Fig. 1) [5, 95]. Depending on the specific process of interest, the appropriate resolution of the model may vary although one may argue that, ultimately, a complete atomistic model with predictive power is the "holy grail."

4.1 Insights from Myosin and AK Studies Regarding Allosteric "Activation Transitions" in Enzymes

In this chapter, using two specific examples, myosin and adenylate kinase, we aim to illustrate how atomistic and CG models can be used to understand better the nature and functional impact of slow "activation motions" in enzymes (Fig. 1). Specifically, we review our studies of the "recovery stroke" in myosin [40–43] and the open/close transition in AK [44–46]. Both transitions occur in the millisecond time scale and are kinetically separable from the chemical step that involves bond breaking/formation in ATP. Therefore these transitions can be studied by various experimental approaches, making it possible to compare computational and experimental results.

The processes that we have analyzed here are allosteric in nature in the sense that they implicate a large number of residues, many of which are far from the active site where substrate binding and chemical transformation occur; they include both domain-scale motions and more localized side chain rearrangements. It is such "multi-scale" nature that makes allosteric transitions both fascinating and challenging to study using computational approaches [96, 97]. The key challenge is to establish among all these multitudes of motions what subset(s) of motions constitute(s) the kinetic bottleneck and what rearrangements have the most significant impact on the subsequent chemical step. A closely related goal is to identify "hotspot" or "hub" residues/interactions that maintain the tight coupling between the multitudes of motions and therefore the long-range and *co-operative* (see below) features of allosteric transitions.

Several approaches have been proposed in the literature to identify "hotspot" residues in allosteric transitions, most notably perhaps being informatics based analysis based on either sequence (e.g., statistical coupling analysis [98, 99]) or structure [100] and various perturbative analyses based on normal mode models of proteins [101, 102]. In our analysis of myosin [41] we found significant overlap between predictions from statistical coupling analysis and hinge analysis based on low frequency normal modes [48]. Mutation of some of these residues was indeed known to lead to the decoupling phenotype in myosin (i.e., decoupling between ATPase and motility), supporting their role in mediating long-range couplings. These results and many recent studies [47, 103–107] have highlighted the functional relevance of intrinsic structural flexibility of proteins, which is often well captured by a small set of low-frequency normal modes or motions prior to the allosteric activation [108, 109]. These observations support the view [96] that many allosteric proteins are constructed from semi-rigid domains or subdomains with hinges and/or semi-rigid subunits, which can move relative to each other, so that the "jigglings and wigglings" (Brownian motion, which is always present at physiological temperatures) can be harnessed through biasing of the free energy surface by ligand binding, modification, and release to propagate the resulting local changes over a long distance to affect activities elsewhere.

A more thorough analysis of "hotspot" residues and kinetic bottleneck of allosteric transitions, however, requires explicitly studying the transition pathway and the corresponding transition state ensemble. This is currently difficult to do with bias-free atomistic simulations for large proteins. As we have illustrated here using myosin and AK, biased atomistic simulations and CG simulations can provide valuable insights, although the results should be interpreted with care considering the approximations inherent in these models. Overall, these more detailed studies provide additional support to the "domain-hinge" view of allostery; on the other hand, they also help highlight additional complexities and various deviations from the somewhat simplistic picture of "domain motions mediated by hinges" [110].

In AK, for example, LID and NMP domain closures are clearly the dominant motions for the open/close processes, and significant closure motions have been observed in both experimental [21] and simulation studies [25–27] even in the absence of the substrates. Mutating the hinge residues such as the unique Pro residues in AKthermo into the corresponding ones in AKmeso indeed induced changes in the domain flexibilities toward AKmeso (especially LID), thus partially supporting the role of hinges in the open/close processes. On the other hand, at least in our simulation models [45], the reverse mutations in AKmeso do not cause significant changes in the LID flexibility, suggesting that hinge flexibility and global transitions are partially decoupled in AKmeso. The limited significance of hinges in AK has also been hinted at by the experimental chimera studies of Bae et al. [24], who found that swapping the entire LID and NMP domain sequences (but not just the CORE-LID hinges) between AKthermo and AKmeso interconverts catalytic properties (and, thus, presumably the open/close rate). Thus the stability and internal flexibility of domains, in addition to the relative displacements of domains, may also contribute to the rate of large scale transitions. Along this line, the "cracking" hypothesis [33, 34, 110] argues that local unfolding/refolding is an important part of allosteric transitions. Our CG studies of AK [44] did not find any compelling evidence for cracking, although more systematic studies are needed before solid conclusions can be drawn; nevertheless, our results do support the important contribution from internal properties of domains, such as dihedral flexibility (Fig. 4b) [44].

Regarding the role of subset(s) of motions in enzyme activation, our studies of the recovery stroke in myosin that integrated QM/MM studies and classical MD simulations serve as a relevant example [42]. By constructing an in silico model based on classical simulation results and explicitly evaluating the ATP hydrolysis barrier as compared to those in various crystal structures, we were able to probe the roles of local and longer-range structural rearrangements in activating the ATPase activity in this prototypical molecular motor. Our results highlight that local changes important to chemistry require stabilization from more extensive structural changes; in this sense, more global structural transitions are needed to activate the chemistry in the active site. In fact, even in more "regular" enzymes that catalyze highly localized chemical transformations, such as hydride transfers, collective structural changes are believed to be important [15]. This is also likely because otherwise active site features conducive to the chemical step are not stably maintained.

It is important to emphasize that our results do not suggest that *all* global conformational changes are required to turn on efficient ATP hydrolysis. In many "decoupling mutants" of myosin [111, 112] the ATPase activity is very close to being normal, indicating that the lack of converter/lever arm rotation does not significantly impair ATP hydrolysis. Therefore, a more likely scenario is a two-phase process (also schematically sketched in Fig. 1): during the first phase, structural transitions near the N-terminus of the relay helix are coupled to the SwII displacement to establish a stable active site and to turn on the ATP hydrolysis; in the second "relaxation phase," the rest of the conformational cascades propagate into the rotations of the converter and the lever arm. This two-phase description is reminiscent of coupled tertiary and quaternary structural changes for O_2-binding-induced allostery in hemoglobin, where it has been shown that tertiary structural changes induced by oxygen binding precede quaternary structural changes [113, 114].

These discussions further highlight the importance of understanding factors that control the coupling between local and global structural changes. If the coupling is very tight such that the various structural rearrangements are highly co-operative, the population of the system in which only a subset(s) of the structural transitions have occurred is extremely low, then it is less meaningful, at least for practical purposes, to ask what subset(s) of motions have a more significant impact on the chemical step; this corresponds to a more "holistic" view of protein structure and motion. In the context of biomolecular motors, ensuring that the transitions between different functional states are highly co-operative is likely essential to maintaining an efficient energy transduction [17, 40], i.e., avoiding wasteful futile ATP hydrolysis that is not tightly coupled to large-scale structural transitions. Whether a high degree of co-operativity is observed in and critical to the function of "regular" enzymes remains to be carefully and systematically analyzed [15]. As to factors (e.g., sequence profile) that control the degree of co-operativity, although "hotspot" residues discussed above are highly relevant, other considerations based on studies of protein folding/stability are expected to be useful as well, especially considering the argument that domain unfolding/refolding might contribute to large-scale transitions in proteins [110, 115].

4.2 *Future Outlooks*

In the following, we briefly comment on several directions that we believe are particularly worthy of further efforts from computational studies, or, more preferably, integrated computational and experimental investigations.

4.2.1 Kinetic Bottlenecks and Co-operativity for Large-Scale Functional Transitions

As emphasized repeatedly in the above discussions, it is important to understand factors that control the kinetics and the degree of co-operativity for large-scale motions in proteins. It is clear that functional transitions often involve both domain-scale motions and local structural rearrangements. Domain motions are more striking in scale while the local transitions more subtle, but the spatial magnitude of changes does not necessarily correlate with kinetic significance. As noted by many studies, large-scale structural transitions are correlated with low-frequency modes, which implies that biomolecules tend to have intrinsic structural flexibilities and the domain-scale motions are largely diffusive in nature; therefore, the kinetic bottleneck of a functional transition may, in fact, consist of key local structural changes that are thermally activated. Moreover, to ensure a high degree of co-operativity, the diffusive domain motions need to be energetically coupled to local rearrangements. Thus it is essential to characterize the underlying free energy landscape of multi-scale motions and the structure-dependent diffusion properties of protein domains (including hydrodynamics effects [68]).

At this moment, with typical computational hardware, characterizing the kinetic bottleneck and the underlying free energy landscape for large systems remains an outstanding challenge [97]. Novel computational techniques and strategies such as transition interface/path sampling [116], milestoning [64], the thermal string method [65], and multi-state Markov state models coupled with massively distributed computing [66, 67] are promising avenues that have mainly been applied to the folding of small peptides/proteins. The milestoning approach has been applied to the recovery stroke of myosin [79], although more in-depth analysis would be useful regarding distinguishing different pathways and the role of key residues. The thermal string approach has also been applied to the transition of ligand (proton) gated ion channels [117] and kinase [118]. Nevertheless, atomic simulations of these sorts remain rather computationally expensive, which makes it difficult to evaluate the convergence and statistical significance of the results. Therefore, at least in the near future, development and validation of effective CG models for proteins remains an attractive and intellectually tantalizing direction [68, 69]. CG models can also be used as a broad approach to sample possible transition pathways, which are then reversely mapped to an atomistic scale and refined [119]. Finally, novel approaches for characterizing and projecting multi-dimensional free energy surfaces have been developed but their application to large protein systems have only just started to appear [120, 121].

4.2.2 Prediction of Functional Transitions at High Resolution

For proteins that undergo large scale conformational transitions, it is not uncommon to have a high-resolution structure for only one of the many functional states (e.g., a molecular motor or kinase with an ATP analog as ligand). Therefore, an important challenge is to predict/construct reliable models for other functional states. Along this line, a productive avenue is to combine computational approaches with low-resolution experimental data for the functional state of interest; good examples include electron microscopy (EM), SAXS, FRET data, and other spectroscopic data that provide structural constraints. For such a purpose, a judicious combination of physical based models (e.g., MD simulations) and structural informatics techniques (e.g., ROSETTA [122] or TASSER [123]) is likely to be most productive in the near future. Notable examples have emerged in recent years, especially concerning using EM and SAXS as low-resolution structural constraints for macromolecular assemblies [124–127]. Further pushing forward the resolution of such models remains a fascinating direction of research.

4.2.3 Connection Between Motions/Dynamics and Enzyme Evolution

The discussions above have been limited to the function of a single protein/enzyme (complex) under in vitro condition. It is tempting to ask how the dynamic properties of proteins/enzymes fit into the broader biological context, such as protein-protein interaction networks in the cell and protein evolution. For example, interesting discussions have been made regarding slow protein conformational fluctuations (i.e., dynamic disorder of protein activity [128]) as an additional origin of stochasticity for the protein interaction network [129]; therefore, the time scale of slow protein fluctuations might need to be tuned in the cellular context to be compatible with the sensitivity and robustness of the underlying protein network. Along this line, it is increasingly realized that the cellular environment is very different from dilute solution and effects such as molecular crowding have a major impact on binding and dynamic properties as well as the stability of biomolecules [130]. Although a significant fraction of the crowding effect can be understood based on "simple" physical arguments such as the excluded volume effect stabilizing more compact conformations relative to extended conformations, to what degree do slow motions of proteins differ (in terms of both time scale and possibly mechanism) between the cellular and in vitro conditions remains to be systematically dissected.

As to the role of motions/dynamics in protein evolution, much consideration has been given to the role of protein flexibility and evolvability [131], and the connection between specific motions and emergence/divergence of function only starts to be explored. As a fascinating example, Johnston et al. [132] reported that a single mutation is able to switch a guanylate kinase enzyme (GK^{enz}) into a functional GK domain (GK^{dom}). This mutation was shown to inhibit the GMP-induced GK domain closure, thus quenching the GK^{enz} activity, but to allow protein binding and spindle orientation, the GK^{dom} function. Although this is an "artificial" example, it is conceivable that modulation of protein motions has played a role in the emergence/divergence of functions during evolution. By combining techniques such as ancestral sequence reconstruction [133] with physical characterization of protein motions, we anticipate that much can be learned regarding how complex motions have been modulated through evolution to help shape the rich functional landscape of proteins in the cell.

Acknowledgments We thank all other collaborators who have also made significant contributions to the studies discussed here. The research has been generously supported by NIH (R01GM071428, R01GM084028 and NLM training grant 5T15LM007359).

References

1. Toscano MD, Woycechowsky KJ, Hilvert D (2007) Angew Chem Int Ed 46:3212–3236
2. Gerlt JA, Babbitt PC (2009) Curr Opin Chem Biol 13:10–18

3. Siegel JB, Zanghellini A, Lovick HM, Kiss G, Lambert AR, Clair JLS, Gallaher JL, Hilvert D, Gelb MH, Stoddard BL, Houk KN, Michael FE, Baker D (2010) Science 329:309–313
4. Hilvert D (2000) Annu Rev Biochem 69:751–793
5. Ma BY, Nussinov R (2010) Curr Opin Chem Biol 14:652–659
6. Boehr DD, McElheny D, Dyson HJ, Wright PE (2006) Science 313:1638–1642
7. Nagel ZD, Klinman JP (2006) Chem Rev 106:3095–3118
8. Boehr DD, Dyson HJ, Wright PE (2006) Chem Rev 106:3055–3079
9. Benkovic SJ, Hammes-Schiffer S (2003) Science 301:1196–1202
10. Pu J, Gao J, Truhlar DG (2006) Chem Rev 106:3140–3169
11. Garcia-Viloca M, Gao J, Karplus M, Truhlar DG (2004) Science 303:186–195
12. Antoniou D, Basner J, Nunez S, Schwartz SD (2006) Chem Rev 106:3170–3187
13. Cui Q, Karplus M (2003) Adv Protein Chem 66:315–372
14. Hammes-Schiffer S, Benkovic SJ (2006) Annu Rev Biochem 75:519–541
15. Hammes GG, Benkovic SJ, Hammes-Schiffer S (2011) Biochemistry 50:10422–10430
16. Hill TL (1977) Free energy transduction in biology. Academic, New York
17. Eisenberg E, Hill TL (1985) Science 227:999–1006
18. Walsh R, Rutland C, Thomas R, Loughna S (2010) Cardiology 115:49–60
19. Kiaris H, Spandidos DA (1995) Int J Oncol 7:413–421
20. Roberts PJ, Der CJ (2007) Oncogene 26:3291–3310
21. Hanson JA, Duderstadt K, Watkins LP, Bhattacharyya S, Brokaw J, Chu JW, Yang H (2007) Proc Natl Acad Sci USA 104:18055–18060
22. Henzler-Wildman KA, Thai V, Ott M, Wolf-Watz M, Fenn T, Pozharski E, Wilson MA, Petsko GA, Karplus M, Hübner CG, Kern D (2007) Nature 450:838–843
23. Henzler-Wildman KA, Lei M, Thai V, Kerns SJ, Karplus M, Kern D (2007) Nature 450:913–916
24. Bae E, Phillips GN Jr (2006) Proc Natl Acad Sci USA 103:2132–2137
25. Arora A, Brooks CL III (2007) Proc Natl Acad Sci USA 104:18496–18501
26. Brokaw JB, Chu JW (2010) Biophys J 99:3420–3429
27. Cukier RI (2009) J Phys Chem B 113:1662–1672
28. Lou HF, Cukier RI (2006) J Phys Chem B 110:24121–24137
29. Kubitzki MB, de Groot BL (2008) Structure 16:1175–1182
30. Pontiggia F, Zen A, Micheletti C (2008) Biophys J 95:5901–5912
31. Beckstein O, Denning EJ, Perilla JR, Woolf TB (2009) J Mol Biol 394:160–176
32. Temiz NA, Meirovitch E, Bahar I (2004) Proteins Struct Funct Bioinf 57:468–480
33. Miyashita O, Onuchic JN, Wolynes PG (2003) Proc Natl Acad Sci USA 100:12570–12575
34. Whitford PC, Miyashita O, Levy Y, Onuchic JN (2007) J Mol Biol 366:1661–1671
35. Bhatt D, Zuckerman DM (2010) J Chem Theory Comput 6:3527–3539
36. Maragakis P, Karplus M (2005) J Mol Biol 352:807–822
37. Lu Q, Wang J (2008) J Am Chem Soc 130:4772–4783
38. Chu JW, Voth GA (2007) Biophys J 93:3860–3871
39. van Wynsberghe AW, Ma L, Chen X, Cui Q (2008) Functional motions in biomolecules: insights from computational studies at multiple scales. In: Schwede T, Peitsch M (eds) Computational structural biology. World Scientific Publishing, Singapore, pp 253–298
40. Yu H, Ma L, Yang Y, Cui Q (2007) PLoS Comput Biol 3:0199
41. Yu H, Ma L, Yang Y, Cui Q (2007) PLoS Comput Biol 3:0214
42. Yang Y, Yu H, Cui Q (2008) J Mol Biol 381:1407–1420
43. Yu H, Yang Y, Ma L, Cui Q (2009) Mechanochemical coupling in molecular motors: insights from molecular simulations of the myosin motor domain. In: Leitner D, Straub JE (eds) Energy flows in proteins. Proteins: Energy, Heat and Signal Flow, CRC Press (2009) pp 23–47
44. Daily MD, Phillips GN Jr, Cui Q (2010) J Mol Biol 400:618–631
45. Daily MD, Phillips GN Jr, Cui Q (2011) PLoS Comput Biol 7:e1002103

46. Daily MD, Makowski L, Phillips GN Jr, Cui Q (2012) Chem Phys 396:84–91, Special issue on "Functional dynamics of proteins"
47. Cui Q, Bahar I (eds) (2005) Normal mode analysis: theory and applications to biological and chemical systems, Mathematical and computational biology series. Chapman and Hall/CRC, New York
48. Li GH, Cui Q (2004) Biophys J 86:743–763
49. Klepeis JL, Lindorff-Larsen K, Dror RO, Shaw DE (2009) Curr Opin Struct Biol 19:120–127
50. Stone JE, Hardy DJ, Ufimtsev IS, Schulten K (2010) J Mol Graph Model 29:116–125
51. Ma J, Flynn TC, Cui Q, Leslie AGW, Walker JE, Karplus M (2002) Structure 10:921–931
52. Schlitter J, Engels M, Krüger P, Jacoby E, Wollmer A (1993) Mol Simul 10:291–308
53. van der Vaart A, Karplus M (2005) J Chem Phys 122:114903
54. Sotomayor M, Schlten K (2007) Science 316:1144–1148
55. Wales DJ (2003) Energy landscapes. Cambridge University Press, Cambridge
56. Huo SH, Straub JE (1997) J Chem Phys 107:5000–5006
57. Ovchinnikov V, Karplus M (2012) J Phys Chem B 116:8584–8603
58. McQuarrie DA (1973) Statistical mechanics. Harper and Row, New York
59. Torrie GM, Valleau JP (1977) J Comput Phys 23:187–199
60. Frenkel D, Smit B (2002) Understanding molecular simulation: from algorithms to applications. Academic, San Diego
61. Barducci A, Bussi G, Parrinello M (2008) Phys Rev Lett 100:020603
62. Ma L, Cui Q (2007) J Am Chem Soc 129:10261–10268
63. Yang S, Banavali NK, Roux B (2009) Proc Natl Acad Sci USA 106:3776–3781
64. Elber R (2011) Curr Opin Struct Biol 21:1–6
65. E W, Vanden-Eijnden E (2010) Annu Rev Phys Chem 61:391–420
66. Bowman GR, Beauchamp KA, Boxer G, Pande VS (2009) J Chem Phys 131:124101
67. Noe F, Schutte C, Vanden-Eijnden E, Reich L, Weikl TR (2009) Proc Natl Acad Sci USA 106:19011–19016
68. Takada S (2012) Curr Opin Struct Biol 22:130–137
69. Tozzini V (2011) Q Rev Biophys 62:333–371
70. Okazaki K, Koga N, Takada S, Onuchic JN, Wolynes PG (2006) Proc Natl Acad Sci USA 103:11844–11849
71. Best RB, Chen Y-G, Hummer G (2005) Structure 13:1755–1763
72. Hills RD Jr, Brooks CL III (2009) Int J Mol Sci 10:889–905
73. Hyeon C, Thirumalai D (2011) Nat Commun 2:487
74. Schliwa M (ed) (2002) Molecular motors. Wiley-VCH, Weinheim
75. Geeves MA, Holmes KC (1999) Annu Rev Biochem 68:687–728
76. Vale RD, Milligan RA (2000) Science 288:88–95
77. Geeves MA, Holmes KC (2005) Adv Protein Chem 71:161–193
78. Fischer S, Windshugel B, Horak D, Holmes KC, Smith JC (2005) Proc Natl Acad Sci USA 102:6873–6878
79. Elber R, West A (2010) Proc Natl Acad Sci USA 107:5001–5005
80. Woo HJ (2007) Biophys Chem 125:127–137
81. Bauer CB, Holden HM, Thoden JB, Smith R, Rayment I (2000) J Biol Chem 275:38494–38499
82. Smith CA, Rayment I (1996) Biochemistry 35:5404–5417
83. Rayment I (1996) J Biol Chem 271:15850–15853
84. Cleland WW, Hengge AC (2006) Chem Rev 106:3252–3278
85. Koppole S, Smith JC, Fischer S (2007) Structure 15:825–837
86. Eisenmesser EZ, Millet O, Labeikovsky W, Korzhnev DM, Wolf-Watz M, Bosco DA, Skalicky JJ, Kay LE, Kern D (2005) Nature 438:117–121
87. Olsson U, Wolf-Watz M (2010) Nat Commun 1. doi:10.1038/ncomms1106
88. Best RB, Hummer G (2005) Proc Natl Acad Sci USA 102:6732–6737
89. Karanicolas J, Brooks CL (2002) Protein Sci 11:2351–2361

90. Zavodszky P, Kardos J, Svingor A, Petsko GA (1998) Proc Natl Acad Sci USA 95:7406–7411
91. Varley PG, Pain RH (1991) J Mol Biol 220:531–538
92. Rundqvist L, Aden J, Sparrman T, Wallgren M, Olsson U, Wolf-Watz M (2009) Biochemistry 48:1911–1927
93. Yang SC, Park S, Makowski L, Roux B (2009) Biophys J 96:4449–4463
94. Turjanski AG, Gutkind JS, Best RB, Hummer G (2008) PLoS Comput Biol 4:e1000060
95. Kamerlin SCL, Warshel A (2010) Proteins Struct Funct Bioinf 78:1339–1375
96. Cui Q, Karplus M (2008) Protein Sci 17:1295–1307
97. Zhuravlel PI, Papoian GA (2010) Q Rev Biophys 43:295–332
98. Lockless SW, Ranganathan R (1999) Science 286:295–299
99. Suel GM, Lockless SW, Wall MA, Ranganathan R (2003) Nat Struct Biol 10:59–69
100. Demerdash ONA, Daily MD, Mitchell JC (2009) PLoS Comput Biol 5:e1000531
101. Zheng WJ, Brooks B (2005) J Mol Biol 346:745–759
102. Balabin IA, Yang WT, Beratan DN (2009) Proc Natl Acad Sci USA 106:14253–14258
103. Bahar I, Rader AJ (2005) Curr Opin Struct Biol 15:586–592
104. Tama F, Brooks CL III (2006) Annu Rev Biophys Biomol Struct 35:115–134
105. Zheng WJ, Brooks BR, Thirumalai D (2006) Proc Natl Acad Sci USA 103:7664–7669
106. Ma JP (2005) Structure 13:373–380
107. Wynsberghe AWV, Cui Q (2006) Structure 14:1647–1653
108. Kern D, Zuiderweg ERP (2003) Curr Opin Struct Biol 13:748–757
109. Gunasekaran K, Ma B, Nussinov R (2004) Proteins Struct Funct Bioinf 57:433–443
110. Whitford PC, Onuchic JN, Wolynes PG (2008) HFSP J 2:61–64
111. Sasaki N, Ohkura R, Sutoh K (2003) Biochemistry 42:90–95
112. Ito K, Uyeda QP, Suzuki Y, Sutoh K, Yamamoto K (2003) J Biol Chem 278:31049–31057
113. Eaton WA, Henry ER, Hofrichter J, Mozzarelli A (1999) Nat Struct Biol 6:351–358
114. Lee AW, Karplus M (1983) Proc Natl Acad Sci USA 80:7055–7059
115. Schrank TP, Bolen DW, Hilser VJ (2009) Proc Natl Acad Sci USA 106:16984–16989
116. Vreede J, Juraszek J, Bolhuis PG (2010) Proc Natl Acad Sci USA 107:2397–2402
117. Zhu FQ, Hummer G (2010) Proc Natl Acad Sci USA 107:19814–19819
118. Gan W, Yang S, Roux B (2009) Biophys J 97:L08–L10
119. Heath AP, Kavraki LE, Clementi C (2007) Proteins Struct Funct Bioinf 68:646–661
120. Das P, Moll M, Stamati H, Kavraki LE, Clementi C (2006) Proc Natl Acad Sci USA 103:9885–9890
121. Gfeller D, De Los Rios P, Caflisch A, Rao F (2007) Proc Natl Acad Sci USA 104:1817–1822
122. Das R, Baker D (2008) Annu Rev Biochem 77:363–382
123. Zhang Y (2008) BMC Bioinf 9:40
124. Topf M, Lasker K, Webb B, Wolfson H, Chiu W, Sali A (2008) Structure 16:295–307
125. Jamros MA, Oliveira LC, Whitford PC, Onuchic JN, Adams JA, Blumenthal DK, Jennings PA (2010) J Biol Chem 285:36121–36128
126. Yang S, Blachowicz L, Makowski L, Roux B (2010) Proc Natl Acad Sci USA 107:15757–15762
127. Trabuco LG, Villa E, Mitra K, Frank J, Schulten K (2008) Structure 16:673–683
128. Lu HP, Xun LY, Xie XS (1998) Science 282:1877–1882
129. Bai F, Wu Z, Jin J, Hochendoner P, Xing J (2012) Slow protein conformational change, allostery and network dynamics. In: Cai W (ed) Protein-protein interactions - computational and experimental tools. InTech, New York
130. Zhou HX, Rivas GN, Minton AP (2008) Annu Rev Biophys 37:375–397
131. Nobuhiko T, Tawfik DS (2009) Science 324:203–207
132. Johnston CA, Whitney DS, Volkman BF, Doe CO, Prehoda KE (2011) Proc Natl Acad Sci USA 108:E973–E978
133. Thornton JW (2004) Nat Rev Genet 5:366–375

Top Curr Chem (2013) 337: 165–188
DOI: 10.1007/128_2012_408

Published online: 19 February 2013

Multiple Intermediates, Diverse Conformations, and Cooperative Conformational Changes Underlie the Catalytic Hydride Transfer Reaction of Dihydrofolate Reductase

Karunesh Arora and Charles L. Brooks III

Abstract It has become increasingly clear that protein motions play an essential role in enzyme catalysis. However, exactly how these motions are related to an enzyme's chemical step is still intensely debated. This chapter examines the possible role of protein motions that display a hierarchy of timescales in enzyme catalysis. The linkage between protein motions and catalysis is investigated in the context of a model enzyme, *E. coli* dihydrofolate reductase (DHFR), that catalyzes the hydride transfer reaction in the conversion of dihydrofolate to tetrahydrofolate. The results of extensive computer simulations probing the protein motions that are manifest during different steps along the turnover cycle of DHFR are summarized. Evidence is presented that the protein motions modulate the catalytic efficacy of DHFR by generating a conformational ensemble conducive to the hydride transfer. The alteration of the equilibrium conformational ensemble rather than any protein dynamical effects is found to be sufficient to explain the rate-diminishing effects of mutation on the kinetics of the enzyme. These data support the view that the protein motions facilitate catalysis by establishing reaction competent conformations of the enzyme, but they do not directly couple to the chemical reaction itself. These findings have broad implications for our understanding of enzyme mechanisms and the design of novel protein catalysts.

Keywords Conformational diffusion · Correlated motions · Dihydrofolate reductase · Hydride transfer barrier · pk_a shift

---

K. Arora and C.L. Brooks III (✉)
Department of Chemistry and Biophysics Program, University of Michigan, Ann Arbor, MI 48109, USA
e-mail: brookscl@umich.edu

Contents

1 Introduction

Enzymes are specialized proteins that catalyze life-sustaining biochemical reactions inside cells. Enzymes permit a reaction that could take years to complete in a solution to be accomplished within a matter of seconds with typically more than a million-fold increase in the reaction rate compared to the uncatalyzed reaction [1, 2]. Deciphering how enzymes achieve such enormous *increase in the rate of the reaction* over similar uncatalyzed reactions in the solution is the Holy Grail of biochemistry. A detailed understanding of enzyme mechanisms can yield great benefits both scientifically and commercially in applications such as the design of biological catalysts and the development of targeted therapeutics to name but a few [3, 4]. Therefore many experimental and theoretical investigations over the decades have been focused on gaining deeper understanding of enzyme mechanisms (cf. [5, 6] and references therein). While these studies have collectively provided significant insights into enzyme function, our understanding of various key mechanisms that enzymes employ to speed up significantly biochemical reactions remains rudimentary [7, 8]. This is evident from our inability to synthesize biological catalysts that can match the efficiency of enzymes produced through natural evolutionary processes [9, 10].

The way an enzyme speeds up a chemical reaction is often a complex, multi-step process. First, the ligand binds to the enzyme. Second, a conformational change occurs that is mostly associated with ligand binding. This transitions the enzyme-substrate complex from an inactive to an active state ("induced-fit" model [11]) or simply stabilizes the pre-existing minor population of the enzyme's active state ("population-shift" model [12, 13]). Third, an actual chemical reaction occurs. Fourth, following chemistry, the product is released and the enzyme returns to its initial substrate-free conformation for the next cycle to begin.

A full and quantitative understanding of the catalytic power of enzymes requires the detailed knowledge of the dynamical changes and the underlying energy landscape (i.e., free energy barriers and minima) as an enzyme progresses through different stages of the catalytic cycle. However, gathering this information is a challenging problem whose solution requires a combination of theory, simulation, and experiment. While the ground state structures combined with kinetic studies provide crucial information on possible mechanisms of enzyme catalysis, they do not unravel atomically detailed transition pathways and the multidimensional energy landscape required to explain the catalytic prowess of enzymes. Theory and simulation can provide complementary information to experiments [14–18]. When data from such simulations is interpreted with a careful attention to the inherent limitations of the methodology, valuable information on atomically detailed conformational transition pathways, reaction intermediates and the *discrete local energy minima of the high dimensional energy landscape of enzyme underlying conformational changes, i.e., conformational substates* [12], can be elucidated to gain a quantitative understanding of enzyme function [19, 20].

This review will present advances in our understanding of enzyme catalysis, primarily exemplified through our group's work in the application of the range of computational methods to investigate quantitatively protein conformational changes and the chemical transformation event in catalysis of a model enzyme dihydrofolate reductase (DHFR) [21–28]. *DHFR catalyzes the reduction of 7,8-dihydrofolate (DHF) to 5,6,7,8- tetrahydrofolate (THF) utilizing the nucleotide cofactor 5,10-methylenetetrahydrofolate reductase (NADPH). The reduction of DHF is initiated through a protonation of the N5 atom of DHF, which is followed by a hydride transfer from the cofactor NADPH to the C6 atom in dihydrofolate.*

Following a brief introduction into DHFR structure and function, we describe key results of equilibrium molecular dynamics (MD) simulations, which provide information on the local conformational fluctuations of the enzyme that may impact the chemical reaction. We then describe insights into factors modulating the chemical reactivity of the enzyme and mediating the rate differences between wild type and debilitating mutants of DHFR obtained through the examination of equilibrium ensembles of activation energies for the key hydride transfer step as well as the distributions of structural parameters in the different protein isoforms. We then discuss insights gained from simulations into the role of protein conformational changes in facilitating the protonation of dihydrofolate that precedes the hydride transfer event in the chemical transformation step. Atomic and energetic details of large-scale conformational reorganization of the enzyme between functional states and the role such conformational changes play in the organization of the reactive groups for efficient catalysis are then described. The features of enzyme catalysis investigated in a model enzyme DHFR may well provide insight regarding the general mechanisms by which enzyme activity is modulated. We note that comprehensive reviews on related topics have appeared that will be of use to interested reader [29, 30]; this brief review makes no attempt at a complete review of the literature.

2 DHFR Structure and Function

DHFR catalyzes the reduction of dihydrofolate (DHF) to tetrahydrofolate (THF) with the help of a cofactor nicotinamide adenine dinucleotide phosphate (NADPH) [31]. The product of this reaction, THF, is a precursor of cofactors required for

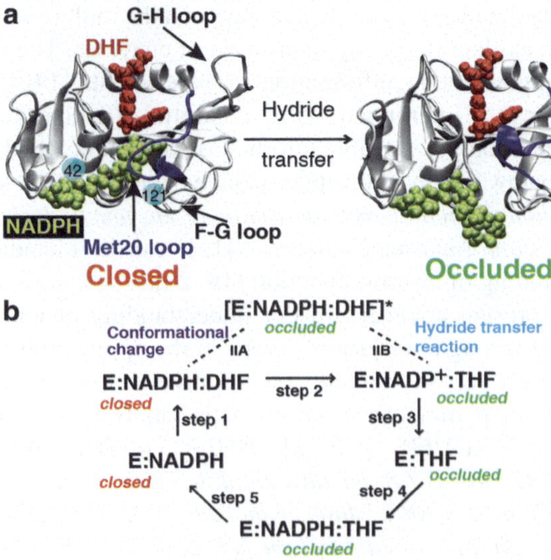

Fig. 1 The structure of *E. coli* DHFR and the conformational changes that take place during the catalytic process. (**a**) Crystal structure of DHFR in the closed (PDB ID: 1RX2) and the occluded (PDB ID: 1RX6*)* states. In the closed state, the Met20 loop stacks against the nicotinamide ring of the cofactor (NADPH) while in the occluded state the loop prevents cofactor from accessing the active site pocket. *Solid balls* illustrate the locations of mutated residues 42 and 121. (**b**) Schematic representation of the catalytic pathway of DHFR as deduced from several kinetic and structural studies. In the holoenzyme E:NADPH and the Michaelis Menten complex E:NADPH:DHF, the Met20 loop adopts the closed conformation. In the three product complexes, E:NADP$^+$, E:THF, and E:NADPH:THF, the Met20 loop occurs in the occluded conformation. We have performed simulations capturing the enzyme's transition between the closed and occluded state (step IIA, above). (Figure 1 of Arora and Brooks [27], Copyright © 2007, The American Chemical Society)

purine, pyrimidine, and amino acid synthesis. Given its biological significance, DHFR has been the subject of several theoretical [21–28, 32–35] and experimental investigations [36–54]. Based on the multiple crystal structures [55] and kinetic data [56] a catalytic cycle for DHFR has been deduced (Fig. 1). These studies suggest that during catalysis, DHFR cycles through five major intermediates: E: NADPH, E:NADPH:DHF, E:NADP$^+$:THF, E:THF, and E:NADPH:THF and undergoes significant conformational changes during this process. Analysis of these structures reveals that the large conformational changes of DHFR are concentrated in its Met20 loop (residues 14–24). Thus, depending upon the conformation of the Met20 loop, different DHFR states along the catalytic pathway are characterized as open, closed, or occluded. In the closed state the Met20 loop stacks against the nicotinamide ring while in the occluded state the Met20 loop sterically hinders the cofactor from binding in the active site (Fig. 1). The conformation of the Met20 loop in turn seems to depend on the ligands bound in the substrate and cofactor binding sites. In the holoenzyme E:NADPH and the Michaelis Menten complex E:NADPH:DHF, the Met20 loop adopts the closed conformation, while in the other three product complexes along the DHFR catalytic pathway, E:NADP$^+$,

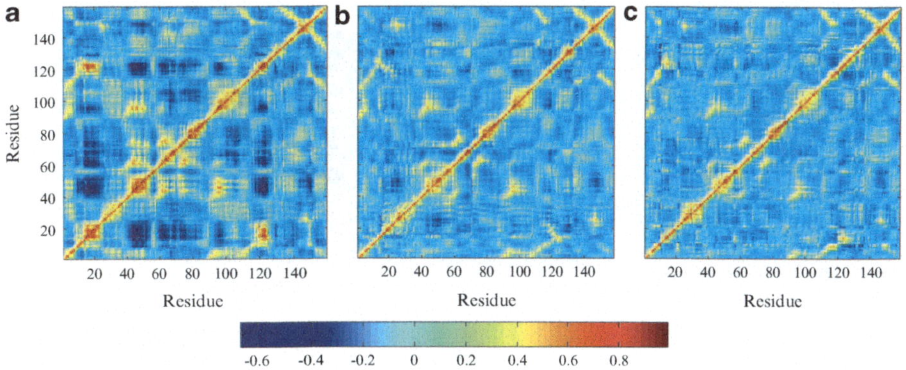

Fig. 2 Residue–residue based map of correlated motions. *Red* and *yellow* indicate regions of positive correlation, and *dark blue* indicates regions of anti-correlation. (**a**) DHFR/DHF/NADPH, (**b**) DHFR/THF/$NADP^+$, (**c**) DHFR/THF/NADPH. (Figure 5 of Radkiewicz and Brooks [25], Copyright © 2003, The American Chemical Society)

E:THF, and E:NADPH:THF, the Met20 loop assumes an occluded conformation. To determine how conformational changes on wide ranging timescales within the enzyme complex relate to its catalytic efficiency, we have generated a conformational and catalytic energy landscape of DHFR. Results of these computational explorations are reviewed in the following sections.

3 Correlated Motion and the Effect of Distal Mutations in Hydride Transfer

Proteins are intrinsically dynamic and protein internal motions can play a key role in their biological function [57, 58]. Initial understanding of the role of such protein motions in DHFR catalysis was derived from the equilibrium molecular dynamics (MD) simulations of wild type DHFR as well as distal mutants of DHFR implicated in modulating the hydride transfer rate by experimental kinetic and mutagenesis studies [32, 42, 59]. Classical MD simulations and correlated motions analysis of the corresponding MD trajectories for three ternary wild type complexes – DHFR/ DHF/ NADPH (DH), DHFR/THF/$NADP^+$ (TP), and DHFR/THF/NADPH (TH) – from the DHFR catalytic cycle have provided residue-based maps of correlated motions [24]. As shown in Fig. 2, the results of correlated motion analysis reveal that spatially and sequentially separated residues in DHFR are coupled, displaying strongly “correlated” motions. However, strongly correlated motions present in the reactant Michaelis complex (DH) structure are abrogated in the product complexes with $NADP^+$ (TP) or NADPH and H4F (TH).

Motivated by earlier investigation of wild type DHFR, MD simulations were performed on reactant Michaelis complexes of G121S, G121V, M42F, and

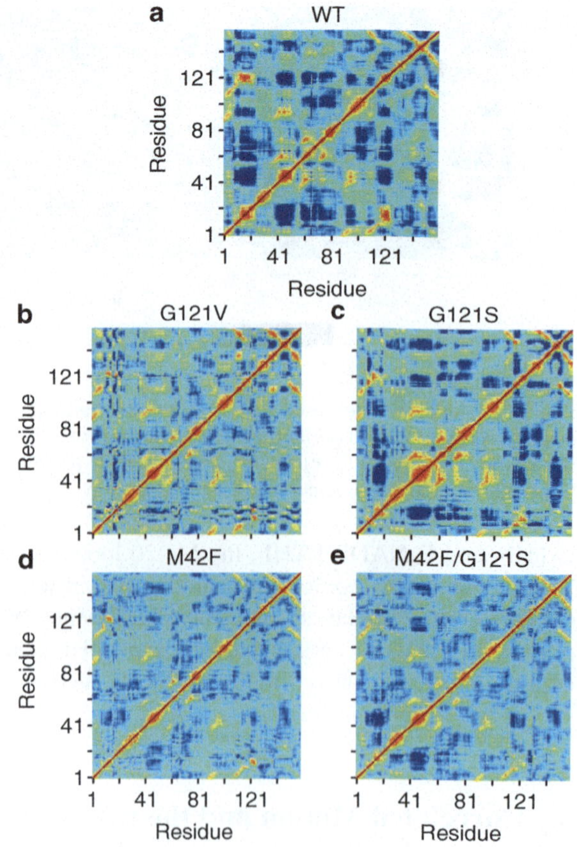

Fig. 3 Covariance matrix for the fluctuations of the C_α atoms in wild type DHFR and various mutants. *Yellow* and *red* regions indicate that the C_α atoms move in a concerted way (positively correlated movements), and *dark blue* means they move opposite to each other (anticorrelated movements). The scale goes from −0.6 (*dark blue*) to 1 (*red*). We note that we get the same qualitative picture for the correlated motions if all heavy atoms are included in the calculation of the covariance matrix. (Figure 3 of Rod et al. [24], Copyright © 2003, National Academy of Sciences USA)

M42F/G121S mutants in which the position of mutations is located far away from the active site [28]. Interestingly, results of this analysis show that compared to wild type DHFR the correlated motions are reduced in all mutant complexes (Fig. 3), correlating somewhat with the reduction in the experimental hydride transfer rates [42]. As shown in Fig. 3, particularly, correlated fluctuations of the Met20 loop with several regions of the enzyme are strongly affected by mutations, suggesting that these motions may be relevant to the catalytic efficacy of the enzyme. The question then arose of if and how diminished correlated fluctuations upon distal mutations would manifest in the structural changes of the enzyme?

Answers were sought by performing cluster analysis of MD trajectories on the Met20 loop backbone conformations based on their ϕ/ψ dihedral angles for G121S, G121V, M42F, and M42F/G121S mutants and wild type DHFR [24]. Specifically, Met20 loop conformations were clustered since the correlated fluctuations of the Met20 loop with the rest of the enzyme are most affected upon mutations (Fig. 4). The results of cluster analysis show that the simulations sample five different well-defined clusters, which include the closed, open, and occluded conformations of the

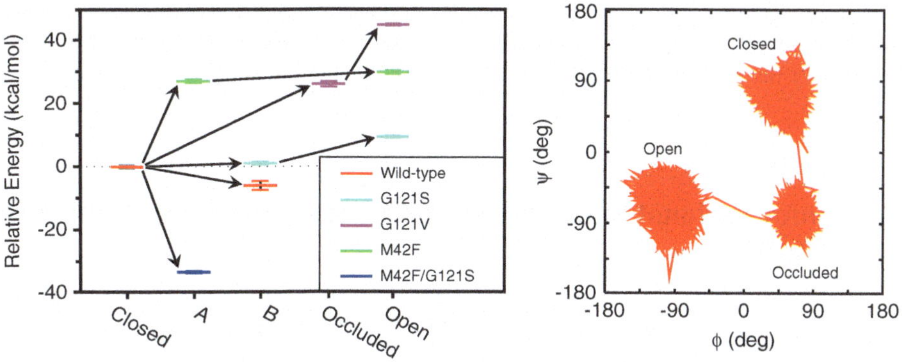

Fig. 4 (*Left*) Energy levels for different conformations of the Met20 loop sampled in the simulations of native and mutant Michaelis complexes of DHFR, relative to the closed conformation. Each *color* represents a particular mutant, and the five different conformations found from our cluster analysis are shown along the horizontal axis. The energies are the average energies calculated by using a generalized Born implicit solvent model from snapshots from the portions of the trajectories belonging to particular loop conformations. The *error bars* represent the standard errors about these averages. The *arrows* indicate the progression in time. (*Right*) Representative trajectory in ϕ–ψ space for loop residue Gly-17 in the G121V mutant to illustrate the extent to which different conformations may be differentiated. (Figure 4 of Rod et al. [24], Copyright © 2003, National Academy of Sciences USA)

Met20 loop. As shown in Fig. 4, the wild type enzyme mostly samples the closed conformation of the Met20 loop. However, the mutants sample multiple conformations of the Met20 loop that include intermediate conformations between crystallographically observed open, closed, and occluded states. Furthermore, results of cluster analysis show that the mutations change the relative energy among the different Met20 loop conformations (Fig. 4). The shift in the energy levels for different conformations of the Met20 loop sampled in the simulations of wild type and mutants indicates that the structural changes occur near the active site, even if the mutation is distant from the site of chemical transformation. These long-range structural perturbations in the vicinity of the active site center of the enzyme may influence the hydride transfer reaction, rationalizing the experimentally observed rate-diminishing effects of distal mutations. How are these structural perturbations manifesting in the energy barrier of the hydride transfer? Answers were sought by performing hybrid quantum mechanics molecular mechanics (QM/MM) simulations capable of capturing bond breaking and formation in the chemical reaction [60, 61], which is beyond the reach of classical MD simulations discussed above. In the next section, we present results of QM/MM investigations of the hydride transfer event in wild type DHFR as well as the G121S and G121V mutants that provide quantitative insights into the affect of mutations outside the active site on the hydride transfer barrier height.

In summary, equilibrium MD simulations of wild type and distal mutants of DHFR show that there are specific correlated motions in the reactive ternary complex of the enzyme that are abolished in the ternary product complex and reduced in the mutants. The changes in the patterns of correlated motions simply reflect structural changes of the catalytically important Met20 loop, which in turn may affect the hydride transfer rate.

4 Conformational Substates Modulate the Barrier to Hydride Transfer

Insights into how structural changes of the Met20 loop can be manifest in the energy barrier of the hydride transfer have emerged from QM/MM calculations of the distributions of activation energies for the hydride transfer reaction in wild type as well as the G121S and G121V variants of DHFR [23]. In these studies, taking into account the possibility that the enzyme may exist as a distribution of conformations [62], the distributions of energy barriers were computed from the corresponding fixed protein structures extracted from the different time-course of MD simulations trajectories. The results of QM/MM simulations show that wide ranges of hydride transfer energy barriers exist for the wild type and the G121V and G121S mutants (see Fig. 5). This result establishes that the energy barrier for hydride transfer and, consequently, reaction rate in DHFR fluctuate in a time-dependent manner. Time-dependent variations in the energy barriers for an enzyme-catalyzed reaction have been demonstrated in both theoretical studies [63] and single molecule experiments [64, 65] by other researchers. In the single molecule experiments, time-dependent variation in the reaction rate was interpreted as a toggling of protein between different conformers, each associated with a distinct reaction rate [64]. However, additional theoretical analyses have revealed that such observations are consistent with the presence of the two or more conformers of a protein with distinct reaction rates [63]. This suggests that the QM/MM calculations of activation energy barriers initiated from several snapshots of the MD simulations are analogous to single molecule experiments, demonstrating time-dependent variation in the reaction rate of individual enzyme molecules [65]. Further, this occurrence suggests that the computationally observed variation in the hydride transfer barriers can be attributed to the existence of the enzyme in multiple distinct conformational substates which modify the potential energy surface, each giving rise to a unique energy barrier for the hydride transfer.

Comparison of the activation energy distributions present in DHFR and its variants demonstrates that the ensemble of energy barriers differs in each system studied (see Fig. 5). *Subsets of structures extracted from MD simulations were employed to calculate the activation free energy for the hydride transfer following the free-energy perturbation approach. These calculations focused on the properties of the reactant states, with an assumption that the corresponding transition states reflect a small perturbation of the reactant-state configurations* [66]. *Furthermore, to account for the inadequate sampling of low hydride-transfer energy barriers at either side of the distribution, Gaussian approximant was used*

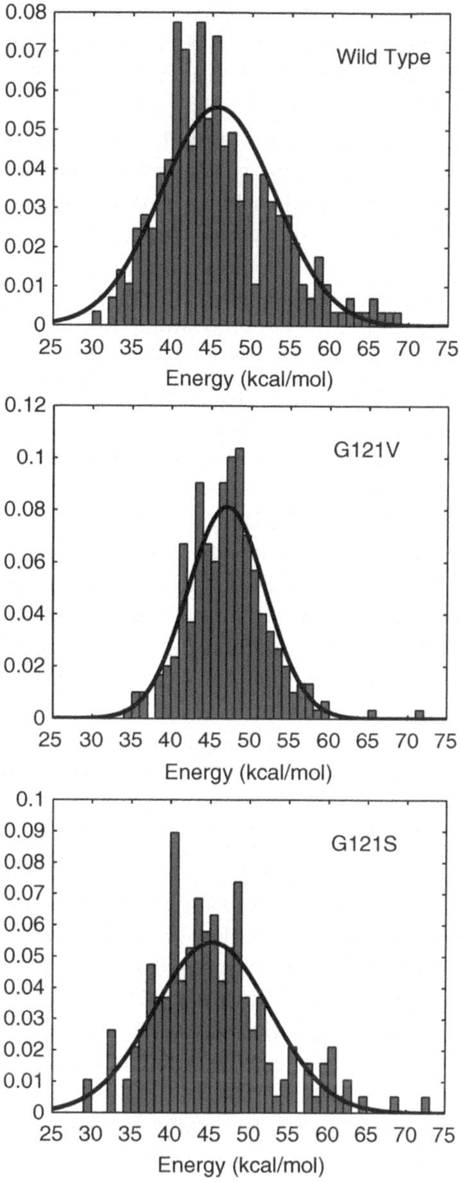

Fig. 5 Histograms of calculated hydride transfer energy barrier distributions for (**a**) wild type, (**b**) G121V, and (**c**) G121S enzymes. The *lines* represent a Gaussian fit to the data. (Figure 2 of Thorpe and Brooks [23], Copyright 2003, © The American Chemical Society)

to model the distribution of the energy barriers for the three enzyme systems. Resulting, effective activation free energies for the wild type, G121S, and G121V distributions are 13.7 (33.4), 12.9 (32.0), and 26.6 (37.6) kcal/mol, respectively [23]. *Values in parentheses correspond to the estimated barriers without Gaussian approximant.* The free energy barrier differences among the three protein systems agree qualitatively with the experimental estimates. The relative ordering of the calculated free energies shows that the energy barrier for the G121V mutant is greater than the wild type, as expected based on experimentally determined rate constants. However, the energy barrier for G121S mutant is not very different from the wild type DHFR. Experimentally, however, the G121S hydride transfer rate differs from that measured for the wild type protein.

How then is the hydride transfer barrier height modulated? To find answers, the hydride barrier distributions of wild type DHFR and its mutants were analyzed for the variability of hydride transfer barriers [22]. The variability of hydride transfer barriers reflects the ground state conformational space explored by an enzyme. In addition, quasi-harmonic (QH) analysis [67] was performed to measure the impact of the Met20 loop conformations on the fluctuations of the ligand and cofactor molecules in the active site. Together these analyses reveal that for the G121S mutant that has a similar free energy as the wild type protein, the variability of hydride transfer barriers is similar to the wild type protein but the flexibility of the ligand and cofactor molecules in the active site is enhanced compared to the wild type protein. This is an unexpected result, as one would anticipate that increased flexibility of the ligands would generate an increase in the variability of the energy barrier distribution. These observations suggest that, although the ligand and cofactor sample more conformational space than in the wild type protein, the additional conformational space explored is not directed toward generating configurations conducive to the hydride transfer, and the corresponding energy barrier distributions are unaffected. Thus, while conformations giving rise to low hydride transfer barriers still exist; such conformations comprise only a small fraction of all conformations accessible to the G121S mutant. This would explain the reduction in the effective hydride transfer rates observed for this mutant. In contrast, for the G121V mutant which has a much higher hydride transfer barrier than the wild type protein, similar analysis shows decreased variability of hydride transfer barriers as well as the reduced flexibility of ligand and cofactor molecules in the active site compared to wild type protein. These observations suggest that the G121V mutant samples very different conformational space from the wild type DHFR and conformations conducive to hydride transfer are not sampled. Taken together, results show that, although there is a subtle difference in the mechanism, eventually for both mutants the decrease in the effective hydride transfer rate arises due to a decrease in the relative amount of conformational substates favorable for hydride transfer to take place compared to the wild type DHFR.

Computational studies by Hammes-Schiffer and coworkers have provided evidence for a network of coupled motions that correlate with the progress of the hydride transfer reaction in DHFR and the G121V mutant [68]. In these studies, thermally averaged geometric properties that may be related to the hydride transfer were computed. We examined the wild type and mutant conformational ensembles for a similar set of representative distances presented by Hammes-Schiffer and coworkers [21–23]. This analysis shows that there are significant differences in the key geometric parameters that correlate well with the progress of the reaction for the wild type and mutant systems. These results highlight differences in the wild type and mutant structural ensembles that may be the origin of the observed

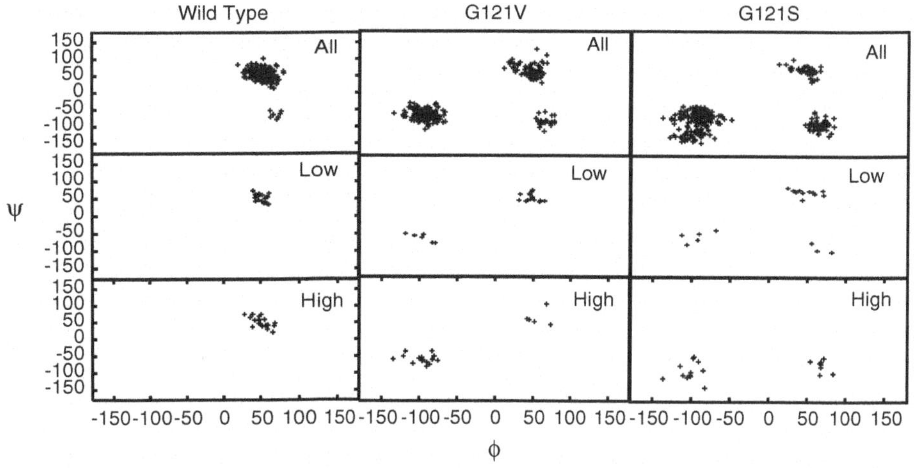

Fig. 6 Backbone ϕ/ψ angles for glutamate 17. The *first row* in each panel corresponds to the entire trajectory, while the *second and third rows* correspond to structures giving rise to low and high transfer barriers, respectively. Dihedral angles are exhibited by the mutant proteins that are not observed in wild type DHFR. These alternate dihedral values are generally associated with higher barriers. Glutamate 17 is observed to be primarily in the region ϕ 50°/ψ 50° for the wild type protein and for the low-barrier structures of the two mutants. Structures giving rise to higher barriers in the mutant proteins are found in the vicinity of ϕ −100°/ψ −50°. (Figure 5 of Thorpe and Brooks [22], Copyright 2004, © The Protein Society)

variation in the distributions of hydride transfer energy barriers and corresponding variations in the reaction rate among the wild type and mutant systems studied. Further, analysis of key geometric parameters reveals that there is a significant qualitative agreement with the earlier study of Hammes-Schiffer and coworkers despite the fact that we employed *equilibrium distributions of static protein conformations* to calculate the hydride transfer barriers. This finding suggests that the changes in the geometric parameters arise due to the progress of the catalytic reaction itself and not due to the fluctuations of the protein. Furthermore, the network of coupled motions observed throughout the protein and ligand by Hammes-Schiffer and coworkers reflect changes in the equilibrium conformational distribution of the protein.

In addition to the key geometric parameters presented by Hammes-Schiffer and coworkers, the detailed examination of the wild type and mutants structural ensembles has revealed a select set of ϕ/ψ dihedral angles that correlate with the presence of protein conformations giving rise to low hydride transfer barriers [22]. As shown in Fig. 6, mutants exhibit dihedral angle values for residues within the Met20 loop not observed in wild type DHFR. These alternate dihedral angle values correspond to the Met20 loop conformations that are not suited for hydride transfer to take place and are generally associated with high hydride transfer energy barriers. These results imply that the configuration of the Met20 loop, which abuts the site of hydride transfer, is one of the important components of a low energy conformational substate. Mutations act to change the structure of this mobile Met20 loop leading to a redistribution of conformational substates present in the protein and thereby impact the hydride transfer rate.

Additional computational studies have emphasized that time-dependent displacement of groups within the active site of the enzyme can directly contribute energy to the reactive event (henceforth referred to as "dynamical coupling"). This phenomenon has been suggested to exist for several enzymatic reactions [69, 70]. However, our results do not support dynamic coupling as a key factor influencing the rate of hydride transfer in DHFR [23]. This is because if the time-dependent conformational fluctuations of the protein groups were directly contributing energy to induce a reactive event, one would expect to see no differences in the energy barriers computed using static snapshots from MD trajectories. Clearly, as discussed above, simulations have demonstrated that differences in the energy barrier distributions exist, even though static snapshots from MD simulations were used for the calculations of the activation energy barriers, suggesting against any role of dynamical coupling in modulating the hydride transfer energy barrier (see Fig. 5). Moreover, differences in the energy barrier distributions that qualitatively agree with the experimentally determined reaction rates suggest that the modulation of the conformational ensemble rather than dynamic coupling is one of the main factors influencing the hydride transfer barriers, and consequently, reaction rate. This suggestion is in agreement with the recent kinetic isotope effect (KIE) study of wild type and N23PP/S148A mutant of DHFR [71]. According to this study, the magnitude and temperature dependence of the KIEs on hydride transfer are unaffected by mutation, suggesting that there is likely no dynamic coupling of protein motions to the hydride transfer step itself.

Other features that could have an impact on the chemical reaction barrier have also been suggested. Theoretical studies by Hammes-Schiffer and coworkers have shown that the protein motions could affect the rate of barrier re-crossing for the hydride transfer step and hence influence the reaction rate [68]. These studies also suggest that, in addition to the effect of protein motions on the barrier re-crossing rate, the incorporation of quantum dynamical effects should further reduce the barrier to hydride transfer by 2–3 kcal/mol [68]. *However, subsequent investigation revealed that both barrier re-crossing and quantum dynamical effects are almost identical for the wild type and mutants suggesting that they are not responsible for differences in the relative reaction rates* [72]. Yet another factor that could influence the reactivity of DHFR is the pK_a of the bound substrate, dihydrofolate [36]. The closed ternary complex can affect hydride transfer, which is believed to follow the substrate protonation at position N5. The substrate protonation is considered to be responsible for the observed pH dependence of the hydride transfer [36]. In the next section we present results of simulations probing the affect of the protein conformational changes in modulating the pK_a of the bound substrate and consequently hydride transfer rate.

In summary, results of QM/MM investigations show that the energy barrier of hydride transfer in wild type G121V and G121S variants fluctuates in a time-dependent manner. The features of these energy barrier distributions are consistent with experimentally determined reaction rates for the three proteins and support observations of single molecule experiments demonstrating time-dependent variation in the reaction rate of individual enzyme molecules. Further, the results show that the fluctuations of hydride transfer energy barriers primarily arise due to the modulation of the equilibrium ensemble of the protein conformations. The configuration of the Met20 loop is an important component of this equilibrium structural ensemble as indicated by the correlation between the change in the dihedral angles populated by residues within the Met20 loop and the concomitant change in the ensemble of hydride transfer barriers. The mutations at position 121 acts to disrupt the equilibrium distributions of the Met20 loop that is adjacent to the site of chemistry in the enzyme and thus impact the hydride transfer rate. Collectively, these results suggest that the alteration of the ensemble of conformational substates populated by the protein rather than dynamical coupling is the key factor influencing the rate constants for the hydride transfer reaction in DHFR.

5 Met20 Loop Facilitates the Protonation of the Substrate

The chemical step of DHFR's catalytic cycle involves a hydride transfer from the cofactor to the substrate with concomitant protonation of N5 atom of the substrate. Experiments have shown that this chemical step is pH dependent [36] and this pH dependency primarily arises due to the substrate protonation that is best described with a single pK_a value of 6.5 [36, 73, 74]. The pK_a value of the substrate in solution is 2.6, implying that DHFR increases the pK_a value by ~4 pK_a units. The key question for investigation is how does an enzyme facilitate the protonation of the substrate?

Answers were sought by performing the pK_a calculations for the key enzyme conformations, those involving the closed and occluded configurations for the Met20 loop [26]. The pK_a values of 7.1 and 7.7 were reported for the closed and occluded Michaelis complexes, respectively. These results confirm that indeed the enzyme facilitates the protonation of the N5 atoms of the substrate. Further, analysis revealed that the side chain of Asp27 in the active site that forms hydrogen bonds to the substrate remains ionized in both occluded and closed complexes to coordinate properly the substrate, thus further stabilizing the substrate protonated state. Together these results suggest that DHFR promotes protonation by enclosing the N5 atom in a hydrophobic pocket together with the negatively charged Asp27 residue. However, in these preliminary investigations the effect of the Met20 loop conformational change in the enzyme on the substrate pK_a and consequently its implications to catalysis were not explored.

Insights into how the conformational change from occluded to closed state enhances the substrate pK_a in the reactive complex have emerged from combined free energy perturbation and molecular dynamics simulations (FEP/MD) for the closed and occluded Michaelis complexes. In this study, initially the flexibility of the Met20 loop was quantified by measuring the C_α root mean square deviation

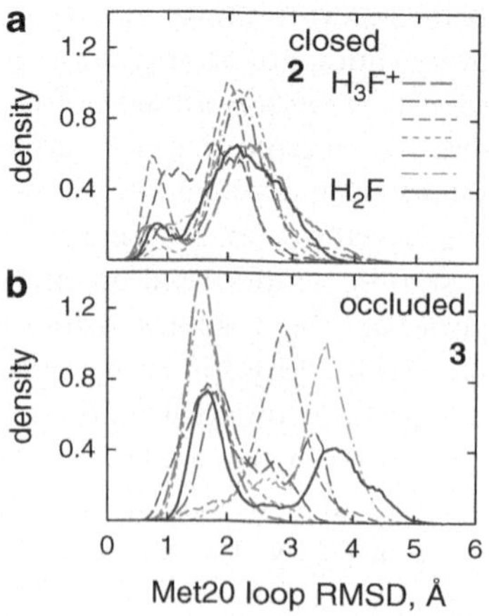

Fig. 7 The distributions of the Met20 loop RMSD in (**a**) the closed complex and (**b**) the occluded complex. All FEP windows are shown (*thin lines*), along with the endpoint states (*thick lines*) to illustrate the overlap between the windows. (Figure 5 of Khavrutskii et al. [26], Copyright © 2007, The Protein Society)

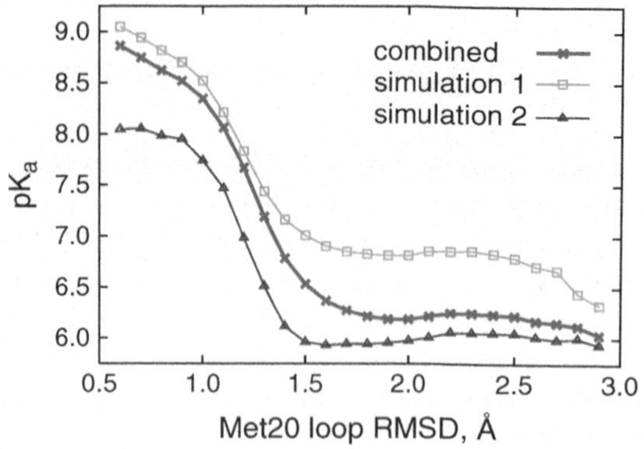

Fig. 8 The substrate pK_a in the closed complex, as a function of the Met20 loop RMSD. The *three curves* refer to the results from the independent simulations 1 (*squares*) and 2 (*triangles*) with different initial velocities and the combined (*crosses*) result. (Figure 6 of Khavrutskiiet al. [26], Copyright © 2007, The Protein Society)

(RMSD) of the Met20 loop with respect to the X-ray closed and occluded complexes. As shown in Fig. 7, the Met20 loop is quite flexible in the occluded complex compared to the closed complex. Furthermore, in the closed complex two major loop states are visible: a small population of "tightly-closed" state (RMSD 0.25–1.0 Å) and a larger population of the "partial closed/open" state (RMSD 1.5–3.5 Å). Subsequently, the Met20 loop flexibility was related to the substrate pK_a by plotting the pK_a of the substrate as a function of the Met20 loop RMSD. As shown in Fig. 8, the pK_a of the Met20 loop strongly modulates the substrate pK_a in the closed Michaelis complex, with the dependence having a characteristic sigmoidal shape. In the "tightly closed" state of closed complex, the computed pK_a is in the 8.0–9.0 unit range that is substantially larger than that in the "partially closed/open" state and the corresponding final pK_a. Further, analysis of the trajectories have revealed that the tight closing of the Met20 loop enhances the interactions of the cofactor and the substrate with the Met20 side chain and aligns the nicotinamide ring of the cofactor coplanar with the pterin ring of the substrate, thus facilitating the protonation.

In summary, results demonstrate that the conformational change of the Met20 loop is coupled to change in the substrate pK_a and may enhance hydride transfer catalysis. These results supplement studies discussed in the previous sections, which show that the equilibrium sampling of the Met20 loop configurations conducive to the hydride transfer is a major factor influencing the catalytic rate.

6 Conformational Dynamics of the Met20 Loop on a Free Energy Surface

Studies discussed so far have suggested that the configuration of the Met20 loop may be an important component of a conformational substate in DHFR that engenders an environment favorable to hydride transfer. However, these conclusions are derived from the equilibrium dynamics simulations that tend to explore localized regions of conformational space around the ground state structures of the enzyme. Therefore a clear link between the conformational reorganization of the Met20 loop between different functional states and the modulation of the chemical environment is missing. This missing dynamics picture has emerged from the elucidation of the conformational transition pathway between closed and occluded states of the enzyme and the corresponding free energy profile using enhanced sampling simulations [27]. These calculations show that the free energy barriers separating occluded and closed states of the Met20 loop in the Michaelis Menten complex of DHFR are small and the transition between these states occurs via an intermediate "open" conformation along the pathway (see Fig. 9). The highest free energy barrier corresponding to this conformational change is 5 $\pm$ 1 kcal/mol. This value of barrier height is much lower than the transition-state theory estimate of 16.0 kcal/mol obtained from the experimental kinetic data [43, 49]. However, the calculated free energy difference between the closed and occluded states (3.3 $\pm$ 1 kcal/mol) is in good agreement with experiments, suggesting that the value for the free energy barrier height obtained from simulations is also probably correct. Moreover, activation barriers estimated by using transition-state theory represent extreme upper limits given the limitation of the theory to describe diffusive processes such as protein folding and protein conformational changes [75].

How does one reconcile the calculated small ~5 kcal/mol free energy barrier with the slow kinetics of the Met20 loop transitions observed in NMR studies? Kramer's reaction rate theory that incorporates the dynamical fluctuations of the enzyme missing in the transition-state theory estimate of rate constant can be used to estimate the reaction rate of diffusive motions in enzymes [76]. Kramer's rate model assumes that the dynamics of the system can be represented as a diffusive process on a low-dimensional free energy surface and has been successfully applied earlier to predict rates of protein folding [77, 78]. To deduce the kinetics of the

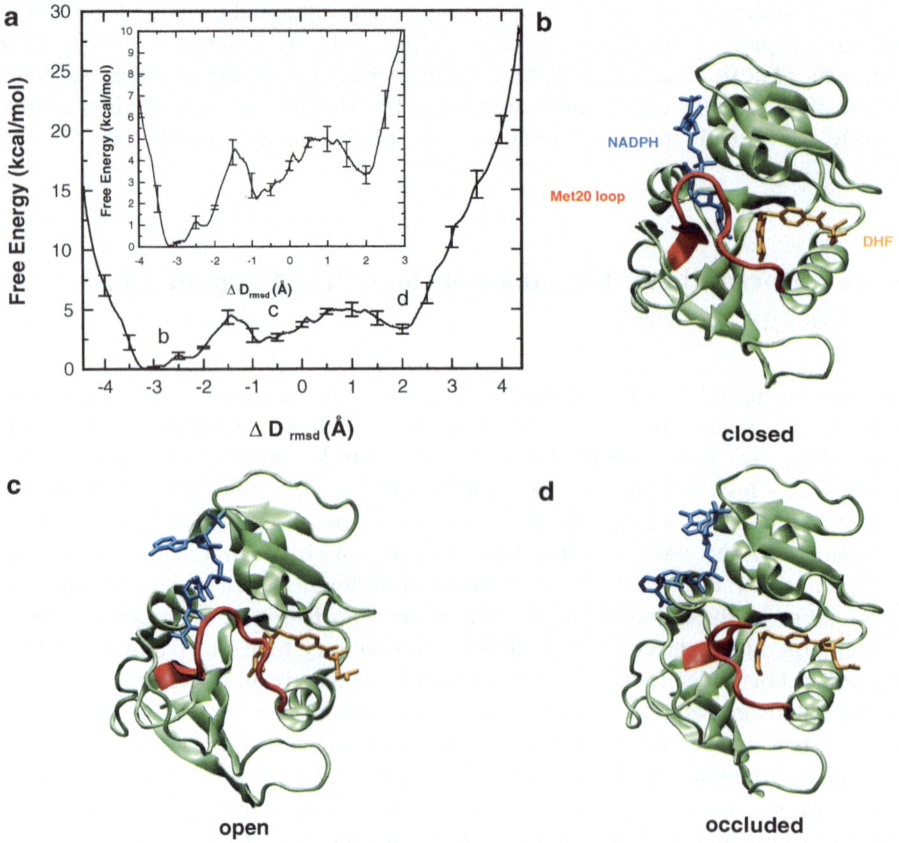

Fig. 9 One-dimensional free energy profile along the ΔD_{rmsd} order parameter. The *error bars* represent the standard deviation in the energy values determined from the set of three umbrella sampling MD simulations with different initial velocities. *Inset* depicts the enlarged view of the same plot. (**a**, **c**, **d**) Representative structures, closed, open, and occluded (*top right* to *bottom right*) corresponding to the three free energy minima (**a**, *top left*) along DHFR's conformational change pathway. (Figure 2 of Arora and Brooks [27], Copyright © 2007, The American Chemical Society)

Met20 loop transitions by using Kramer's reaction rate theory, the position dependent diffusion constants along the Met20 loop conformational transition pathway were computed. The computed diffusion constant values together with the barrier height obtained from the free energy profile were then incorporated into Kramer's reaction rate equation. Interestingly, the calculated rate of transition from the closed state using Kramer's rate relationship agrees very well with the rate of the Met20 loop transitions provided by NMR dispersion experiments [47]. Thus, simulations show that the slow dynamics of the system arises due to the small diffusion constant on a rugged energy landscape and not due to high-energy barriers separating different conformational states.

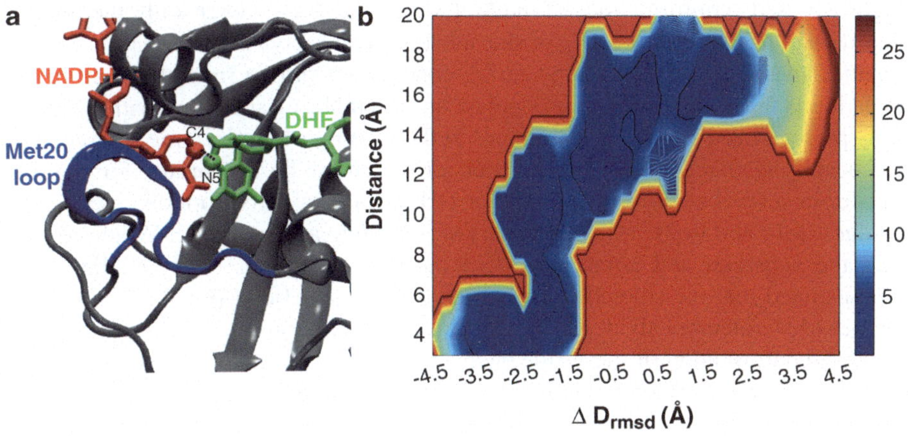

Fig. 10 (**a**) View of the active site and Met20 loop corresponding to the closed state depicting atoms involved in the hydride transfer reaction. (**b**) The two-dimensional free energy surface along the ΔD_{rmsd} order parameter vs hydride transfer distance between the atom N5 of DHF and C4 of NADPH. In the closed state ($\Delta D_{\mathrm{rmsd}} \approx -3.5$ Å), the distance between the cofactor and substrate fluctuates in the range of 3–4 Å, suitable for promotion of the hydride transfer reaction. (Figure 3 of Arora and Brooks [27], Copyright © 2007, The American Chemical Society)

The small free energy barriers separating the functionally important conformational states imply that the system can populate alternate conformations via thermal fluctuations. As shown in Fig. 9, when the cofactor is out of the binding pocket, the enzyme can frequently sample open and occluded conformations because of a small (~3 kcal/mol) free energy barrier between the two states. However, when the cofactor is in the binding pocket, the closed conformation is thermodynamically most favored [27]. These results suggest that every possible conformation of the protein is present at all times, although with different population distributions that can be modulated via interaction with the specific substrate. This mechanistic picture emerging from the thermodynamic and kinetic analysis of the Met20 loop fluctuations is consistent with the population-shift model of ligand binding, which postulates that the ligand binding merely stabilizes the pre-existing minor population of the enzyme's active state in the conformational ensemble [13]. This viewpoint is also consistent with the perspective of conformational dynamics in DHFR informed by NMR dispersion experiments, according to which each intermediate in the catalytic cycle samples low-lying excited states whose conformations resemble the ground state structures of the preceding and following intermediates [49].

The detailed examination of the closed state conformational ensemble has revealed that only a few selected conformations of the Met20 loop from the ensemble have an active site geometry conducive to the hydride transfer reaction (see Fig. 10). This phenomenon was noticed in the previous studies of this system. As discussed in Sect. 5, simulations exploring the role of the Met20 loop conformations in modulating the substrate protonation have shown that in the closed state two substates can be distinguished, namely the "partially closed" and "tightly-closed" conformations of the Met20 loop [26]. However, only the tightly-closed state has the correct pK_a of the substrate conducive to protonation. Taken together, these results reaffirm the earlier conclusions drawn from the equilibrium simulations that the enzyme can populate an ensemble of substates with differentially preorganized protein environments in the vicinity of the enzyme's active site.

Local side chain motions have been proposed to be important for hydride transfer in several enzymes [79, 80]. For DHFR catalysis, hydrophobic active site residues Ile14 and Ile94, which in the crystal structure are in van der Waals contact with the substrate and cofactor, are of particular interest. NMR studies have demonstrated that the side chains of residues Ile14 and Ile94 populate both *trans* and *gauche*$^+$ rotamers about the χ_1 dihedral angle in solution. However, only the *gauche*$^+$ conformation of these side chains is observed in the occluded and closed crystal structures. Further, modeling has suggested that in the *trans* rotameric state, the side chains of these residues would sterically clash with the atoms of the cofactor and pterin rings, respectively. Therefore in order for residue Ile14 to exist as a *trans* rotamer, the nicotinamide ring would have to be displaced towards the pterin ring. To test this hypothesis, we computed the free energy surfaces corresponding to the χ_1 dihedral angle of the residues Ile14 and Ile94 along the reaction coordinate, as part of the conformational change of the Met20 loop [27]. As shown in Fig. 11, the *trans* rotamer population is observed only in the open state and in high-energy conformations leading to the occluded state of that loop. Only the *gauche*$^+$ and a small amount of *gauche*$^-$ populations are present in the closed, reaction competent state. Interestingly, the position of the *trans* population along the reaction coordinate coincides with a decrease in the hydride transfer distance. These simulation results are consistent with the hypothesis that favors a mechanism in which residues Ile14 and Ile94 guide the cofactor and substrate toward a reactive configuration for subsequent hydride transfer reaction and thus facilitate catalysis. Corroborating these findings, recent studies examining the relationship between the hydride-donor acceptor distance (DAD) and its distribution and dynamics to the rate of hydride transfer and the temperature dependence of intrinsic KIEs has provided evidence that residue Ile14 participates in the restrictive active-site motions that modulate the DAD and thus assist the hydride transfer [81].

In summary, characterization of microsecond-millisecond timescale conformational fluctuations that precede the chemical transformation step of DHFR show that the largest conformational changes are concentrated in the Met20 loop of the enzyme implicated in modulating the hydride transfer efficacy. The free energy profile corresponding to the conformational transition pathway reveals that the free barriers separating the functional states of enzyme are small and rough, suggesting rapid sampling of conformers. Furthermore, results show that the slow kinetics of the enzyme suggested by NMR dispersion experiments arises due to the diffusive motions of the Met20 loop on the rough energy landscape rather than the large reorganization energy barrier as suggested by the application of transition-state theory to estimate barrier heights from the experimental kinetics data. Finally, these studies also emphasize the role of the assembly of enzyme side chains to achieve optimal geometry for chemistry.

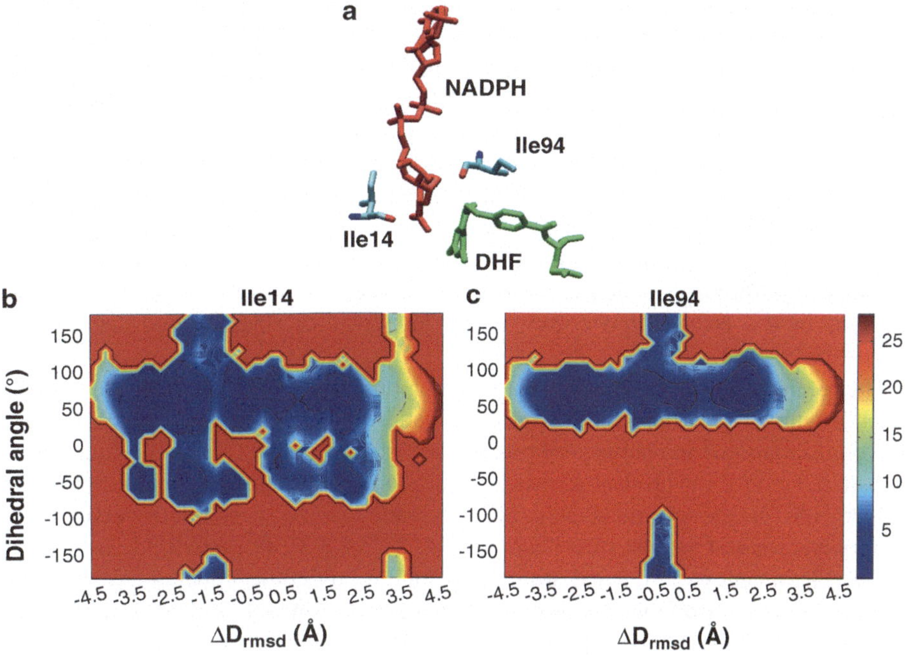

Fig. 11 (**a**) View of the active site showing the side chains of residues Ile14 and Ile94 near the cofactor and the substrate, respectively. (**b**, **c**) Two-dimensional free energy surface corresponding to the χ_1 rotameric dihedral angle (N-C_α-C_β-$C_{\gamma 1}$) along the reaction coordinate for residues Ile14 (*middle*) and Ile94 (*right*). Both residues explore a minor *trans* conformation near the barrier region (see Fig. 9). In the *trans* rotameric state these residues interact unfavorably with the substrate and cofactor bringing the two reactants close to each other. (Figure 4 of Arora and Brooks [27], Copyright © 2007, The American Chemical Society)

7 Concluding Remarks

The simulations discussed above have provided detailed quantitative insights into the DHFR reaction mechanism not currently accessible to experiments. As demonstrated, equilibrium MD simulations of DHFR show that there are specific correlated motions in the reactive ternary complex of the enzyme that are abrogated in the ternary product complex and reduced in the mutants, suggesting that these motions are relevant to the catalytic efficacy of the enzyme. These correlated fluctuations of the enzyme manifest as the distinct conformational substates that are correlated with modulations in the hydride transfer barrier height as indicated by the examination of equilibrium ensembles of activation energies for the key hydride transfer step as well as the distributions of structural parameters in the different protein isoforms. Backbone conformations of the Met20 loop are important components of these conformational substates as illustrated by the link between dihedral angle values exhibited by residues within the Met20 loop and modulation of the hydride transfer barrier height. The backbone ϕ/ψ dihedral angles of wild type DHFR predominantly occupy regions of conformational space that are conducive to hydride transfer, while the mutant proteins visit alternative conformations generally associated with high barriers. Clearly the mutations act to disrupt the equilibrium distribution of the Met20 loop that abuts the site of chemistry. This alteration of the static, equilibrium distribution of the Met20 loop conformations is sufficient to explain the influence of mutations on the rate of the hydride transfer. The above discussion clearly suggests that the Met20 loop of DHFR serves as a master regulator of catalysis due to its ability to adapt to mutations by assuming alternate conformations that lead to redistribution of conformational substates present in the protein and consequently give rise to distinct hydride transfer barriers.

In recent papers, the question of whether protein motions are coupled to the chemical transformation has been hotly debated. Although it is broadly accepted that the conformational movements of proteins are indispensable to enzyme function in substrate binding and product release, the question of whether protein motions play a direct role in the chemical step of enzymatic catalysis is controversial with arguments for and against such coupling [34, 51, 82, 83]. For example, in the case of DHFR, recent experimental studies reported a "dynamic knockout" caused by mutations, suggesting that the conformational fluctuations can influence the chemical step of enzyme catalysis [51]. In these studies, mutations were made that both prevent formation of the occluded conformation through loss of hydrogen bonding between the Met20 and GH loops of DHFR and impair millisecond timescale motions of the Met20 loop in the Michaelis complex. This N23PP/S148A DHFR mutant displayed a reduced rate constant for hydride transfer compared to wild type and it was suggested that this was a consequence of the loss of conformational flexibility. However, subsequent computational [34] and experimental [71] KIE studies of the N23PP/S148A variant of DHFR reached quite the opposite conclusion to that reached by the previous investigation [51]. In the latter studies it was suggested that the role of flexibility in catalysis is negligible and that the largest contribution to catalysis comes from the pre-organization of the active site, which is a longstanding concept [84, 85] that has been implicated in function of other enzymes [86]. Consistent with these suggestions, based on the understanding of DHFR's reaction mechanism emerging from our own computational investigations as discussed above, we suggest that the decrease in the hydride transfer rate constant of N23PP/S148A variant likely arises due to modulation of the equilibrium conformational ensemble as the result of the structural perturbations to the reactive distance in the active site.

In summary, our picture of the enzyme mechanism, as assembled from the quantitative investigation of conformational changes and chemical transformation of DHFR, reveals that alteration of conformational equilibrium rather than dynamical coupling is the key factor influencing the rate of the reaction. Protein fluctuations are important in establishing the structural ensembles conducive to the chemical reaction but do not directly couple to the chemical reaction itself. *This view of enzyme catalysis is also supported by other calculations and experiments and has been discussed elsewhere in the literature* [29, 87]. Finally, we conclude by mentioning that DHFR is an amazing molecular machine, in which conformational heterogeneity, cooperative conformational changes, and multiple intermediates are integral to its catalytic function.

Acknowledgments This work was supported by the National Institutes of Health through the Center for Multi-Scale Modeling Tools for Structural Biology (grant RR012255) and the National Science Foundation through the Center for Theoretical Biological Physics (PHY0216576).

References

1. Fersht A (1985) Enzyme structure and mechanism. W. H. Freeman, San Francisco
2. Jencks WP (1987) Catalysis in chemistry and enzymology. Dover Publications, Mineola, New York
3. Liang JF, Li YT, Yang VC (2000) J Pharm Sci 89:979
4. Pollard DJ, Woodley JM (2007) Trends Biotechnol 25:66
5. Benkovic SJ, Hammes-Schiffer S (2003) Science 301:1196
6. Garcia-Viloca M, Gao J, Karplus M, Truhlar DG (2004) Science 303:186
7. O'Brien PJ, Hollfelder F (2010) Curr Opin Chem Biol 14:634
8. Nagel ZD, Klinman JP (2009) Nat Chem Biol 5:543
9. Röthlisberger D, Khersonsky O, Wollacott AM, Jiang L, DeChancie J, Betker J, Gallaher JL, Althoff EA, Zanghellini A, Dym O, Albeck S, Houk KN, Tawfik DS, Baker D (2008) Nature 453:190
10. Frushicheva MP, Cao J, Chu ZT, Warshel A (2010) Proc Natl Acad Sci USA 107:16869
11. Koshland DE (1994) Angew Chem Int Ed Engl 33:2375
12. Frauenfelder H, Sligar SG, Wolynes PG (1991) Science 254:1598
13. Gunasekaran K, Ma B, Nussinov R (2004) Proteins 57:433
14. Karplus M, Kuryian J (2005) Proc Natl Acad Sci USA 102:6679
15. Warshel A (1989) Computer modeling of chemical reactions in enzymes and solution. Wiley, New York
16. Arora K, Brooks CL III (2011) In: Frank J (ed) Molecular machines in biology. Cambridge University Press, Cambridge, p 59
17. Karplus M, McCammon JA (2002) Nat Struct Biol 9:788
18. Schlick T, Collepardo-Guevara R, Halvorsen LA, Jung S, Xiao X (2011) Q Rev Biophys 44:191
19. Arora K, Brooks CL III (2007) Proc Natl Acad Sci USA 104:18496
20. Radhakrishnan R, Schlick T (2004) Proc Natl Acad Sci USA 101:5970
21. Thorpe IF, Brooks CL III (2005) J Am Chem Soc 127:12997
22. Thorpe IF, Brooks CL III (2004) Proteins Struct Funct Bioinf 57:444
23. Thorpe IF, Brooks CL III (2003) J Phys Chem B 107:14042
24. Rod TH, Radkiewicz JL, Brooks CL III (2003) Proc Natl Acad Sci USA 100:6980
25. Radkiewicz JL, Brooks CL III (2000) J Am Chem Soc 122:225
26. Khavrutskii IV, Price DJ, Lee J, Brooks CL III (2007) Protein Sci 16:1087
27. Arora K, Brooks CL III (2009) J Am Chem Soc 131:5642
28. Rod TH, Brooks CL III (2003) J Am Chem Soc 125:8718
29. Hammes GG, Benkovic SJ, Hammes-Schiffer S (2011) Biochemistry 50:10422
30. Nagel ZD, Klinman JP (2010) Chem Rev 110:PR41
31. Blakely RL (1984) In: Blakely RL, Benkovic SJ (eds), Folate and Pteridines, vol 3. Wiley, New York, p 191

32. Agarwal PK, Billeter SR, Rajagopalan PTR, Benkovic SJ, Hammes-Schiffer S (2002) Proc Natl Acad Sci USA 99:2794
33. Chen J, Dima RI, Thirumalai D (2007) J Mol Biol 374:250
34. Adamczyk AJ, Cao J, Kamerlin SCL, Warshel A (2011) Proc Natl Acad Sci USA 108:14115
35. Liu H, Warshel A (2007) Biochemistry 46:6011
36. Fierke CA, Johnson KA, Benkovic SJ (1987) Biochemistry 26:4085
37. Cannon WR, Singleton SF, Benkovic SJ (1996) Nat Struct Biol 3:821
38. Epstein DM, Benkovic SJ, Wright PE (1995) Biochemistry 34:11037
39. Miller GP, Benkovic SJ (1998) Biochemistry 37:6327
40. Miller GP, Benkovic SJ (1998) Biochemistry 37:6336
41. Miller GP, Wahnon DC, Benkovic SJ (2001) Biochemistry 40:867
42. Rajagopalan PTR, Lutz S, Benkovic SJ (2002) Biochemistry 41:12618
43. McElheny D, Schnell JR, Lansing JC, Dyson HJ, Wright PE (2005) Proc Natl Acad Sci USA 102:5032
44. Schnell JR, Dyson HJ, Wright PE (2004) Biochemistry 43:374
45. Schnell JR, Dyson HJ, Wright PE (2004) Annu Rev Biophys Biomol Struct 33:119
46. Osborne MJ, Schnell J, Benkovic SJ, Dyson HJ, Wright PE (2001) Biochemistry 40:9846
47. Boehr DD, McElheny D, Dyson HJ, Wright PE (2010) Proc Natl Acad Sci USA 107:1373
48. Boehr DD, Dyson HJ, Wright PE (2008) Biochemistry 47:9227
49. Boehr DD, McElheny D, Dyson HJ, Wright PE (2006) Science 313:1638
50. Venkitakrishnan RP, Zaborowski E, McElheny D, Benkovic SJ, Dyson HJ, Wright PE (2004) Biochemistry 43:16046
51. Bhabha G, Lee J, Ekiert DC, Gam J, Wilson IA, Dyson HJ, Benkovic SJ, Wright PE (2011) Science 332:234
52. Oyeyemi OA, Sours KM, Lee T, Kohen A, Resing KA, Ahn NG, Klinman JP (2011) Biochemistry 50:8251
53. Nagel ZD, Dong M, Bahnson BJ, Klinman JP (2011) Proc Natl Acad Sci USA 108:10520
54. Oyeyemi OA, Sours KM, Lee T, Resing KA, Ahn NG, Klinman JP (2010) Proc Natl Acad Sci USA 107:10074
55. Sawaya MR, Kraut J (1997) Biochemistry 36:586
56. Benkovic SJ, Fierke CA, Naylor AM (1988) Science 239:1105
57. Tzeng SR, Kalodimos CG (2012) Nature 488:236
58. Frederick KK, Marlow MS, Valentine KG, Wand AJ (2007) Nature 448:325
59. Cameron CE, Benkovic SJ (1997) Biochemistry 36:15792
60. Warshel A, Levitt M (1976) J Mol Biol 103:227
61. Gao JL, Truhlar DG (2002) Annu Rev Phys Chem 53:467
62. Frauenfelder H, Parak F, Young RD (1988) Annu Rev Biophys Biophys Chem 17:451
63. Zhang Y, Kua J, McCammon JA (2003) J Phys Chem B 107:4459
64. Lu HP, Xun L, Xie XS (1998) Science 282:1877
65. Zhang Z, Rajagopalan PTR, Selzer T, Benkovic SJ, Hammes GG (2004) Proc Natl Acad Sci USA 101:2764
66. Tobias DJ, Brooks CL III (1987) Chem Phys Lett 142:472
67. Brooks CL III, Karplus M, Pettitt BM (1988) Proteins: a theoretical perspective of dynamics, structure, and thermodynamics. Wiley, New York
68. Agarwal PK, Billeter SR, Hammes-Schiffer S (2002) J Phys Chem B 106:3283
69. Knapp MJ, Klinman JP (2002) Eur J Biochem 269:3113
70. Antoniou D, Caratzoulas S, Kalyanaraman C, Mincer JS, Schwartz SD (2002) Eur J Biochem 269:3103
71. Loveridge EJ, Behiry EM, Guo J, Allemann RK (2012) Nat Chem 4:292
72. Watney JB, Agarwal PK, Hammes-Schiffer S (2003) J Am Chem Soc 125:3745
73. Morrison JF, Stone SR (1988) Biochemistry 27:5499
74. Beard WA, Appleman JR, Delcamp TJ, Freisheim JH (1989) J Biol Chem 264:9391
75. Ansari A, Jones C, Henry E, Hofrichter J, Eaton W (1992) Science 256:1796

76. Hanggi P, Talkner P, Borkovec M (1990) Rev Mod Phys 62:251
77. Best RB, Hummer G (2006) Phys Rev Lett 96:228104
78. Bryngelson JD, Onuchic JN, Socci ND, Wolynes PG (1995) Proteins 21:167
79. Schwartz SD, Basner JE (2004) J Phys Chem B 108:444
80. Hay S, Pudney CR, Sutcliffe MJ, Scrutton NS (2008) Angew Chem Int Ed 47:537
81. Stojković V, Perissinotti LL, Willmer D, Benkovic SJ, Kohen A (2011) J Am Chem Soc 134:1738
82. Pisliakov AV, Cao J, Kamerlin SCL, Warshel A (2009) Proc Natl Acad Sci USA 106:17359
83. Kamerlin SCL, Warshel A (2010) Proteins Struct Funct Bioinf 78:1339
84. Warshel A (1978) Proc Natl Acad Sci USA 75:5250
85. Warshel A, Sharma PK, Kato M, Xiang Y, Liu H, Olsson MHM (2006) Chem Rev 106:3210
86. Olsson MHM, Warshel A (2004) J Am Chem Soc 126:15167
87. Benkovic SJ, Hammes-Schiffer S (2006) Science 312:208

Top Curr Chem (2013) 337: 189–208
DOI: 10.1007/128_2012_412

Published online: 19 March 2013

Protein Dynamics and the Enzymatic Reaction Coordinate

Steven D. Schwartz

Abstract This chapter discusses progress over the past 15 years in understanding the role of protein dynamics in enzymatically catalyzed chemical reactions. Research has shown that protein motion on all timescales from femtoseconds to milliseconds can contribute to function, and in particular in some enzymes there are sub-picosecond motions, on the same timescale as barrier passage, the couple directly to chemical transformation, and are thus part of the reaction coordinate. Approaches such as transition path sampling and committor analysis have greatly enhanced our understanding of these processes.

Contents

S.D. Schwartz (✉)
Department of Chemistry, University of Arizona, 1306 East University Blvd., Tucson, AZ 85721, USA
e-mail: steve.schwartz@einstein.yu.edu

1 Introduction

Fifteen years ago we suggested [1] that a possible explanation for anomalous experimental enzyme kinetics may be direct coupling of picosecond protein dynamics to reaction. The last 15 years have seen this suggestion – and in fact the general proposal that enzyme dynamics is important in enzyme chemistry – debated with unusual fervor. While the importance of protein motion to such phenomena as ligand rebinding in myoglobin [2] has been acknowledged and seemingly accepted for almost 40 years, the suggestion we made 15 years ago is different for two reasons. First, ligand binding, while a chemical event, does not involve change in chemical identity. This is a difference that is psychological, but the alteration in language associated with chemistry has proven significant. Second, the experimental claims made for ligand binding relate to conformational heterogeneity, implying that transitions between conformations are slow compared to the time to rebind, and so static on the timescale of the event of interest. Theoretical models can then be built from direct averaging over rate constants that are indexed by conformation [3]. We have proposed that protein motions on all timescales ranging from those equivalent to barrier passage (or penetration) to slow conformational rearrangement are *potentially* important for the chemical event in enzymes. These differences in focus between enzymatic chemistry and ligand binding have resulted in a significantly larger level of debate and, as such, this debate has focused the science very effectively. This focus has lent significantly greater clarity to the issues in the last few years, and these investigations will be the topic of this review.

A good part of the argument that has arisen around the subject of protein dynamics and enzymatic function has been driven by semantics, and so we now wish to state clearly and carefully the subject of this investigation. We quite simply wish to understand how enzymes work. From a chemist's perspective, this means we want to understand in atomic detail how chemical reaction occurs at an enzyme active site. It is our goal to investigate this question in a completely unbiased fashion. The only "modeling" that has been used in the investigations we have reported come from the use of widely accepted numerical tools such as the CHARMM [4] molecular mechanics potential augmented by QM/MM technologies [5, 6] at the reaction center. Our stated goal is to make no assumption as to mechanism, but rather to allow mechanism to be discovered in the process of the computations we undertake. This mechanism is described by the "reaction coordinate," in other words the set of all atomic motions necessary to allow reaction to occur. Given the complexity of chemical reaction in enzymatic systems, special techniques are needed, and they will be described in this review.

We will review studies of three separate enzyme systems (lactate dehydrogenase (LDH), purine nucleoside phosphorylase (PNP), and dihydrofolate reductase (DHFR)). Most of our results will be theoretical but in cases in which direct collaborations have resulted in experiment we will report those as well. The end result is that, in both LDH and PNP, there is direct coupling of protein motion on the timescale of barrier passage as part of the basic mechanism or reaction coordinate. Further, in LDH we have shown evidence for the fact that this motion, or promoting vibration, is created by the protein architecture itself. In other words the motion that corresponds to the promoting vibration is not simply part of the reaction coordinate but is a unique direction in the protein structure. In DHFR, this is not in fact the case. There is no organized promoting vibration. We will discuss how these results fit with other results in the literature, both for specific enzymes and in the general framework of rate theories.

2 Dynamics in Enzymes and the Reaction Coordinate: The Picosecond Timescale

Since our suggestion 15 years ago that protein dynamics could play a role in passage over the chemical barrier in enzymatic systems, a continuing feature of interest to us and consternation to others has been the fact that the protein motion is quite rapid when considered relative to normal enzymatic turnover rates. Most enzymes have a turnover rate of about 1 ms^{-1}. The motions we have discovered – promoting vibrations – have been in the hundreds of femtoseconds to picoseconds timescale. In fact, several other investigators over the years have now found other enzymes in which such rapid promoting vibrations are of import [7–11]. There is of course no contradiction in this disparity of timescale; the rate-limiting step for many enzymes is in fact product release or substrate binding. What this implies is that the enzyme spends the vast majority of the time from its initial encounter with substrate until product release "engaged in activities" other than barrier passage (to be discussed in a later section). In the following section of this review we discuss our results describing these rapid promoting vibrations. In particular we find that two of three enzymes of very different chemistry employ such a mechanism (LDH and PNP) and one (DHFR) which, catalyzing the same chemistry as LDH, appears to not support a promoting vibration. The section starts with a brief description of the computational methodologies we employ. Should this not be of interest to the reader, the section can be skipped.

2.1 *Methodologies*

In order to understand rigorously the process of chemical reaction in an enzyme using theoretical methods we need to follow the atomic motions in the entire system as reaction occurs. Because chemical reaction in an enzyme is an extremely rare event, simply running trajectories using quantum mechanics to determine reactive potential energy surfaces (QM/MM techniques for example) will not yield the desired information. What is needed is a rare event simulation methodology. We choose to use the transition path sampling (TPS) [12–16] methodology of Chandler and co-workers. The critical idea behind this advance was to find "reactive" trajectories for complex phenomena via a Monte Carlo walk in trajectory space. (The word reactive is in quotes because it certainly can be applied to situations in which chemical reaction is not involved [17].) Gathering a large number of these trajectories allows further calculation of an object known as the stochastic separatrix [18]. The stochastic separatrix is the statistical analogue of the transition state or, more precisely, the transition state ensemble. It is the collection of points in conformation space where initiation of trajectories with random momenta chosen from a Boltzmann distribution will result in 50% of the trajectories proceeding to products and 50% returning to reactants.

In fact, this analysis of the stochastic separatrix provides methods for testing a reaction coordinate. Motion along the stochastic separatrix is motion that does not progress the reaction to either products or reactants. Thus, a putative reaction coordinate can be tested by restraining the reaction coordinate degrees of freedom while all other degrees of freedom are allowed to evolve dynamically. If the choice of the degrees of freedom is correct, then unrestrained trajectories will yield a distribution of probabilities to react strongly peaked about five. Initial methods of reaction coordinate identification involved such a "guess and check" approach. We have developed a method based on the structure of the separatrix itself [19, 20]. This method is highly successful and extremely efficient, yielding a good representation of the reaction coordinate "for free" when the separatrix is calculated. The method is distinct from the committor distribution fitting methods that have been developed [21–24] for the determination of reaction coordinates. The fitting methods are best applied to diffusive barrier passage, whereas our methods are best applied to direct barrier passage, the kind most often encountered in enzymatic chemical reactions. Our definition of the components of the reaction coordinate is simply those degrees of freedom that do not change on the separatrix. In this sense then, the separatrix must be narrow in these degrees of freedom. The problem is that we do not have a mathematical description of the separatrix, only points in space. The separatrix is likely "narrow" in these special degrees of freedom in some non-Cartesian coordinate system. Thus what is needed is a method that identifies the correct parameterization of the separatrix and in doing so finds the coordinates for which there is minimal change. Finding the correct coordinate system in which to describe the separatrix is a nonlinear fitting problem, and we use methods from machine learning known as kernel PCA [25]. A usable representation of the topology of the separatrix is obtained when a single nonlinear component is shown to reproduce accurately the entire separatrix. The members of the reaction coordinate are then identified as those that make minimal contribution to this single nonlinear principle component. We have shown [19, 20] that the components found closely align with those we "guessed and checked" in the past. The identification of reaction coordinates will likely continue to be one of the most challenging areas in rare event simulation. We note that it is critical that any proposed reaction

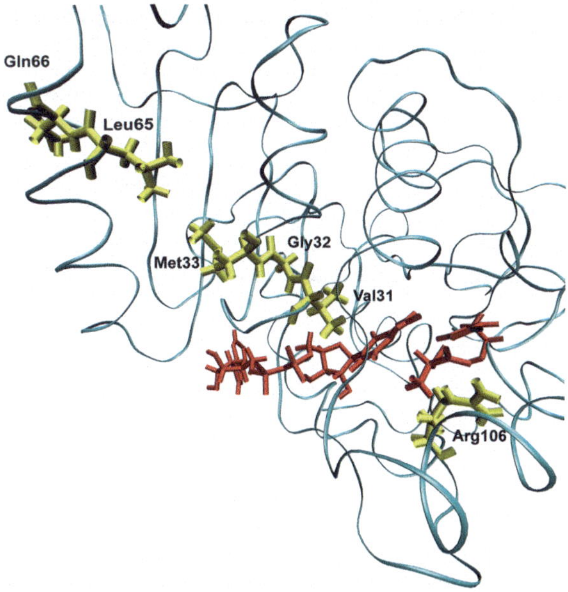

Fig. 1 Ribbon diagram of a monomer of human heart lactate dehydrogenase demonstrating the promoting vibration. The donor and acceptor, the nicotinamide ring, and lactate are colored *red*. The promoting vibration residues, Val31, Gly32, Met33, Leu65, Gln66 (*yellow*) are compressed towards the active site bringing the NC4 of the nicotinamide ring and substrate carbon closer together while Arg106 (*yellow*) relaxes away locking the substrate in product formation. These residues span the entire length of the monomer to the edge of the protein. Reprinted with permission from J Chem Phys B 10.1021/jp207876k 2011

coordinate be checked with the committor distribution method we have described. This method, first suggested by Onsager [18] and later expanded upon by Chandler and co-workers [12–16], demonstrates that motion does not just accompany reaction but is in fact a causal necessity for reaction.

2.2 LDH

The first enzyme to which we applied the TPS methodology was human heart lactate dehydrogenase [26, 27]. A large number of reactive trajectories were gathered and were tested for decorrelation. Examination of the trajectories made it immediately clear that human heart LDH has as part of each reactive trajectory a compressive motion that extends across the body of the protein to cause the donor and acceptor to approach closely. This motion is shown graphically in Fig. 1. Testing showed that this motion is an obligate part of the reaction coordinate. The timescale of this motion is roughly 150 fs. We thus now know that there is motion in the body of the protein that brings the donor and acceptor sites in LDH

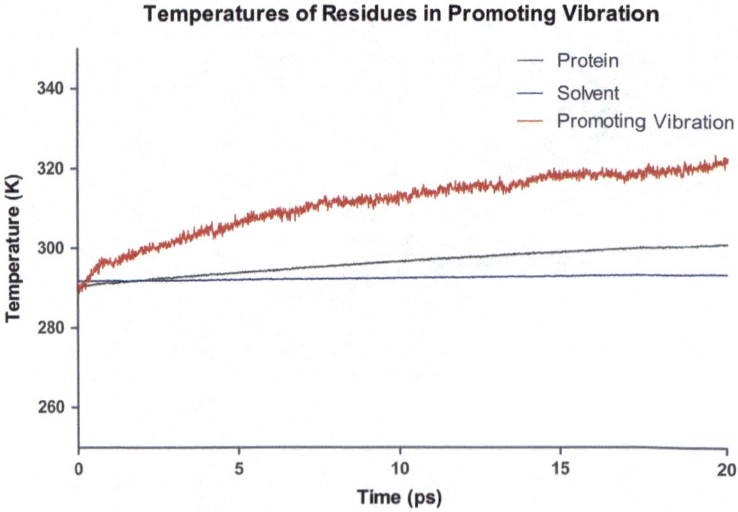

Fig. 2 Computational heating experiment of the active site of LDH. The nicotinamide ring is heated and the temperature of the solvent, protein and residues along the promoting vibration are monitored. The promoting vibration is clearly hotter than the rest of the protein indicating a preferred direction of thermal energy transfer. Reprinted with permission from J Chem Phys B 10.1021/jp207876k 2011

closer together. This promoting vibration, however, need not be any specially selected motion in the vibrations of the protein matrix. It could simply be the case that there are all types of diffusive heat transfer in the protein, and it just so happens that this specific mode is effective in contributing to catalysis. This is exactly the way proton transfer in solution occurs. In a completely stochastic way, donor and acceptor are caged and randomly brought into close enough proximity that reaction can occur. The question then is whether all vibrational motions in the protein are "created equal," or whether the promoting vibration is in some way a special outgrowth of the architecture of the protein. In some sense this is not important to understanding how the chemistry works, but it is critically important to developing an understanding of how the enzyme is constructed.

Recent work [28] in our group has addressed exactly this question. In order to investigate this question we performed a numerical experiment much like that which Straub has implemented for myoglobin [29–31]. We pumped heat into the active site and monitored the temperature at residues in the promoting vibration and the rest of the protein. Figure 2 shows the result of the temperature of the solvent, the protein as a whole, and that of the promoting vibration. It is clear from these data that not only is the promoting vibration unique in terms of chemistry; it is also unique in terms of protein architecture. Thermal energy transferred from the active site is preferentially channeled through the promoting vibration. In fact, in this work we were also able to show that in concentric shells in the protein, the residue(s) of the promoting vibration are statistically "hotter" than almost all other residues a

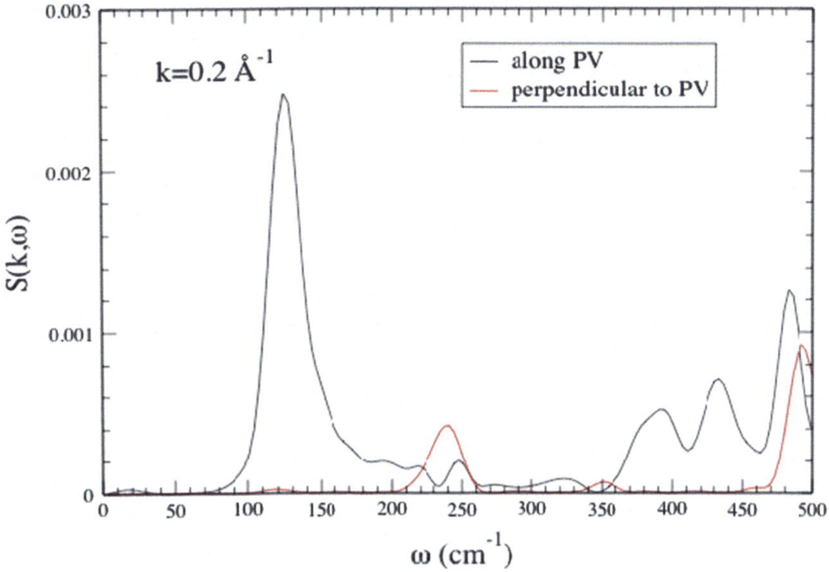

Fig. 3 The structure factor $S(k, \omega)$ for $k = 0.2$ Å^{-1} along the PV axis (*black line*) and along an axis perpendicular to the promoting vibration. There is strong anisotropy. The sharpness of the peak at about 150 cm^{-1} means that there are stable fluctuations along the PV axis for that frequency. Reprinted with permission from J Chem Phys B 10.1021/jp210347h 2011

similar distance from the active site. In addition, it is worth mentioning that these calculations provide a perfect experimental test for the existence of a promoting vibration. Laser spectroscopic methods are able to monitor such temperature differences [32].

These results show that the promoting vibration in LDH is in fact a unique direction in the protein backbone, but it really does not inform directly on the mechanism for promoting vibration transferring of local heat energy from the exterior to the active site. In other words, when thermal fluctuations occur on the surface of the protein, are they preferentially channeled inward in the same way that the simulated laser radiation is channeled outward? To address this question we computed the dynamic structure factor both orthogonal to and along the direction of the promoting vibration. This quantity – which is the experimental observable in, say, a neutron scattering experiment – is the Fourier transform of the density–density correlation function, and along a given wavevector $\vec{k}$, and it describes time dependent mass fluctuation. A peak at a given frequency component indicates that along that wavevector there is significant mass movement. Representative results are shown in Fig. 3. It is clear that in the range of 150 wavenumbers there is a significant peak in the direction of the promoting vibration showing that density fluctuations in the appropriate time regime are located along the promoting vibration axis. This shows that the promoting vibration is in fact a "soft" direction in the protein, and vibrational energy is easily transmitted along it. There is no obvious reason for this physical property to be developed in the protein, but speculation is possible. It is certainly a possibility that the promoting vibration was in some way a selected characteristic favored by evolution. Another equally plausible (perhaps more so) explanation is this "soft" direction is a part of the protein folding pathway for the particular enzyme, and it has been co-opted for chemistry.

2.3 PNP

Though the dehydrogenases/reductases are central to metabolism, we must ask whether the hydrogen transfer reactions are unique in the involvement of protein dynamics in chemical barrier passage. We thus became interested in seeing whether a promoting vibration was present in a very different class of enzymatic chemistry. We studied the reaction catalyzed by purine nucleoside phosphorylase: scission of the ribosidic bond in nucleosides. From a variety of experimental evidence we had reason to believe that this may well be the case. First, mutation on the surface of this protein that had absolutely no effect on the structure of the active site of the enzyme increases the rate of on enzyme chemistry by about a factor of 2 [33]. In addition, crystal structures of transition state inhibitors showed a unique stacking of electronegative oxygen atoms directly adjacent to the scissile ribosidic bond. Our hypothesis was that compression of this bond could sufficiently polarize the electron density so as to make the base a better leaving group. A variety of computations [34–36] confirmed this hypothesis. The reaction mechanism from an atomistic point of view was also seen to be more complex than that of the hydride transfer of LDH. Figure 4 shows the difference between a committor distribution for a complex reaction like PNP (in the top panel) and one for a direct mechanism like that for LDH (bottom panel). It is clear that more time is spent in the transition region for PNP. It should, however, be emphasized that this "long" residence time is still measured in femtoseconds! The PNP reaction is actually a two step process, with the first rise in committor coming from a promoting vibration compression and initial breaking of the ribosidic bond and the second rise coming from the relative motion of the ribose ring towards the phosphate to form the ribose-1-phosphate product.

2.4 DHFR

There remains controversy in the field as to whether promoting vibrations on this timescale are important in what might be termed the "hydrogen atom" of hydride transfer: dihydrofolate reductase. There is a suggestion in work of Hammes-Schiffer and colleagues [37] that in fact faster motions accompany reaction. What is not studied in their computations is the central issue of causality. That is, do motions of protein atoms simply happen at the same time as reaction at the active site, or are they in fact causal of reaction. In order to answer this question, we have

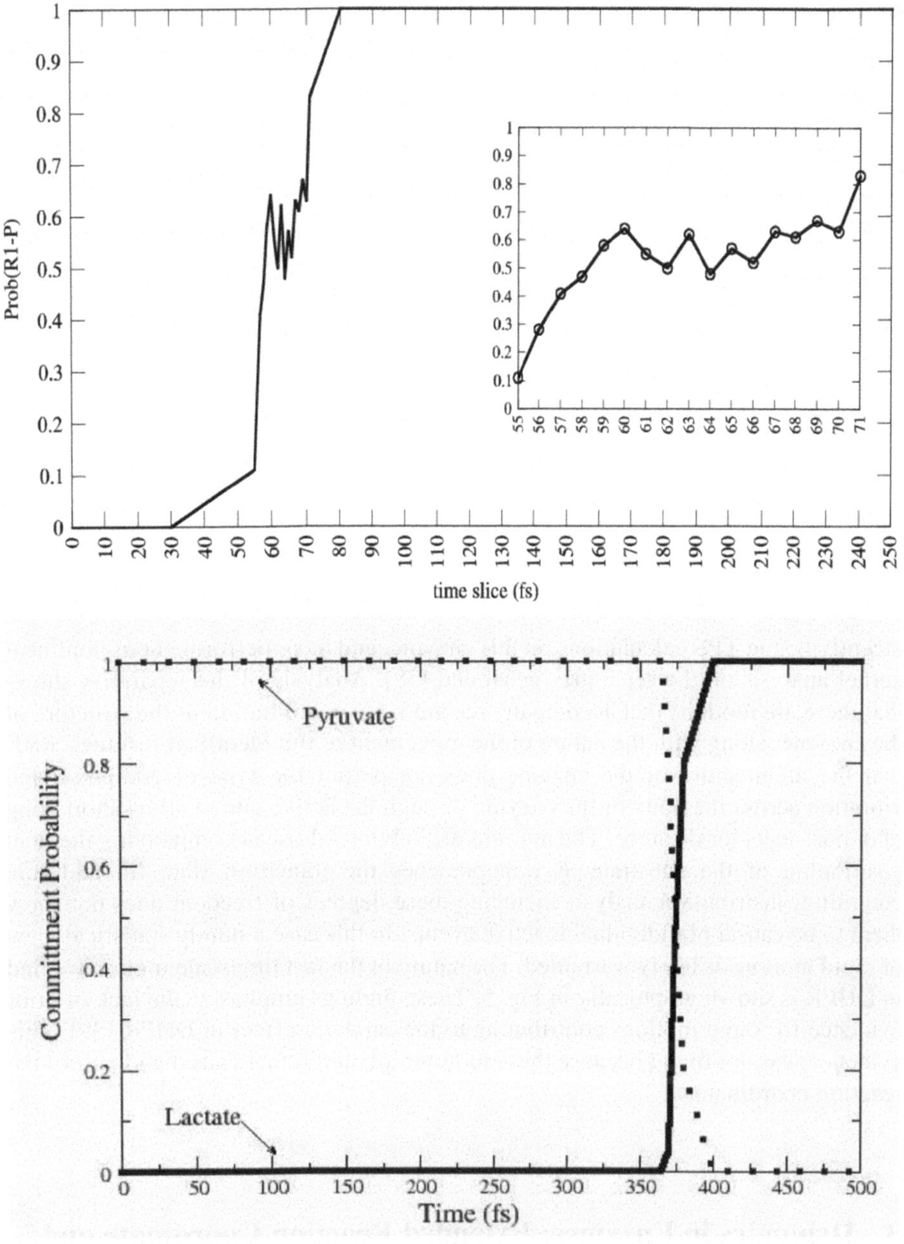

Fig. 4 Commitment probability values were calculated for both PNP (*top panel*) and LDH (*lower panel*) along sample trajectories from the reactive path ensembles for the respective enzymes. It is clear that the transition region for LDH is far shorter in time duration than for PNP, but even for PNP the residence time is on the order of 10–20 fs. Reprinted with permission from J Chem Phys B 10.1021/jp207876k 2011

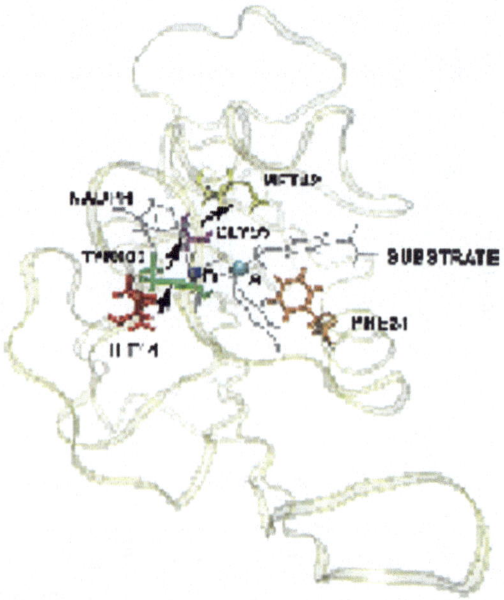

Fig. 5 Representation of the direction of motion of residues found in DHFR to be concurrent with the reaction. D is the donor and A is the acceptor. The substrate and the NADPH molecules are shown in *gray* and the rest of the protein is shown as *ribbons*. Subsequent committor analysis showed that this motion was not in fact part of the reaction coordinate, and so not a causal part of the reaction mechanism

recently begun TPS calculations on this enzyme, and have performed our nonlinear kernel analysis on the separatrix generated [38]. Analysis of the separatrix shows that there are motions that accompany reaction, but examination of the structure of the enzyme, along with the nature of the movement of the identified residues, finds that the architecture of the enzyme does not permit the type of compressional vibration across the body of the enzyme through the active site to allow shortening of donor–acceptor distance. The motions are subtler – those accompanying the final positioning of the substrate as it approaches the transition state. In addition, committor distribution analysis including these degrees of freedom does not show them to be causal of individual reactive events. In this case a purely statistical view of rapid motions is likely warranted. The nature of the fast timescale motion we find in DHFR is shown graphically in Fig. 5. These findings emphasize the lack of prior evidence for rapid motions contributing to the catalytic effect in DHFR [39]. This evidence was not found because these motions are not in fact causal and part of the reaction coordinate.

3 Dynamics in Enzymes: Extended Reaction Coordinate and Conformational Fluctuations

As was described above, the importance of conformational distributions has long been understood and accepted in enzymatic function. Hammes-Schiffer's view of coupled motions (see for example [40]) fits within this realm. The idea here is simply that transitions between states is slow on the timescale of barrier passage. What we wish to do now is, first, understand what those transitions actually are and, second, develop an understanding of how these transitions relate to the rapid motions we have already identified in LDH and PNP. In addition, since it is obvious that actual barrier passage is quite rapid (femtoseconds to picoseconds), but the time between substrate binding and product release is many order of magnitude slower, we need to find what might actually be happening in a single instance of the enzyme during this process. Another way of describing this is we wish to extend the reaction coordinate back from the barrier closer to substrate binding, at least in a general statistical sense. The timescales are simply far too long just to continue classical trajectories and, even if we could complete a single classical trajectory, the more important investigation is a statistical study of multiple conformations and their interconversions, not the traversal of a single reactive trajectory. We again first describe the methodologies we employ. As in the previous section, this can be skipped if one wishes to avoid the technical details.

3.1 Methodologies

The first technical difficulty presented by this endeavor is that the timescales are simply too long for convention molecular dynamics. Obviously one cannot simply take the reactive trajectories we have generated and extend them back in time long enough to view the earlier processes that occur in a catalytic event. In some fashion, then, a method is needed to search the highly complex free energy surface of the protein. This then highlights the second technical difficulty, which is that this free energy surface is almost assuredly quite complex with a great many local minima. If one simply performed a Monte Carlo walk on the surface with annealing to local minima, one would find only a tiny fraction of the possible minima, and those may well not all be functionally relevant. The goal is to search the surface in such a fashion that the chemically distinct basins [41] are explored in at least a reasonable fashion.

Our method to accomplish this grew out of the recognition that, in the reaction catalyzed by LDH, the distance of the donor and acceptor is central to reactivity. Thus, if one wishes to locate the functionally important conformations, they must all have the donor–acceptor distance within a range that allows reaction [42]. The specific computational tool we choose was to restrain this distance to that found in the crystal structure, and allow the rest of the protein to explore conformational space via Langevin dynamics and minimization [43]. Once a set of candidate conformations were obtained in this fashion, they were further minimized with gradual release of the donor–acceptor constraints. Cases in which the donor–acceptor distance remained short were deemed catalytically competent, and those in which it grew beyond a specified cutoff were deemed to be not catalytically competent. Once these conformations were obtained, we wish to get some physical picture of how they interconvert. Again, because interconversion is a very rare event, and in fact is far too long for simple molecular dynamics, we are forced to develop a multistep process. First, a minimum free energy path between the conformations is found using a temperature dependent string method [44, 45]. Once this minimum free energy path is located (assumed to dominate conformation interconversion) the actual free energy along this physical coordinate (labeled by RMSD from the starting structure [46]) can be found by umbrella sampling. Finally, using a simplified reaction coordinate, we can employ QM/MM sampling to obtain the chemical barrier along the conformational coordinate. These methods have been described in detail [43].

3.2 Conformational Fluctuations in LDH

The first observation that we made was that, in the vast majority of cases, release of the constraints allows the donor and acceptor to move far beyond the selected range and so to fall into conformations that are not catalytically competent. In fact, the number was far less than 10%. We have now studied this process on both sides of the chemical barrier in the reaction, and have obtained similar results. We next wanted to understand the energetics of conformational motion in the enzyme. To do this we found the minimum free energy path between all sets of basins that had low donor–acceptor distance. The minimum average potential energy path and the associated free energies we discovered on a selected path are shown in Fig. 6. Discovery of these abstract paths is worthwhile, but the real goal of this investigation is developing an understanding of the types of physical changes the enzyme undergoes in these longer time interconversions. These basins have relatively small but detectable differences in geometry of the active site, and this is shown graphically in Fig. 7. The central difference between the various low donor–acceptor distance conformations was in the positioning of the active site loop over the reactive center. It is known that reaction cannot occur in LDH if this loop is not closed, and this result shows how critical positioning of this loop is to catalysis. In fact, analysis of the paths *between* the short donor–acceptor conformations both always involve a significant widening of the donor–acceptor distance, and partial opening of the loop conformation. The effective free energy barrier to these motions from the free energy minimum (that is not reactive) was always in the 10 kcal/mol range. This is simply a statement of the rarity of these catalytically competent conformations. In addition, we have shown that at the conformational free energy maxima the short donor–acceptor distance has the lowest chemical barrier to reaction – almost zero.

Thus to summarize, we have found the nature of motions and conformations that bring the system to the catalytically competent state that have short donor–acceptor distance. These conformations are very rare and there is a free energy barrier of entropic origin toward these competent conformations. This last finding suggests that perhaps an appropriate description of catalysis in this enzyme is a search for these rare competent conformations, followed by the reactive event that is assisted

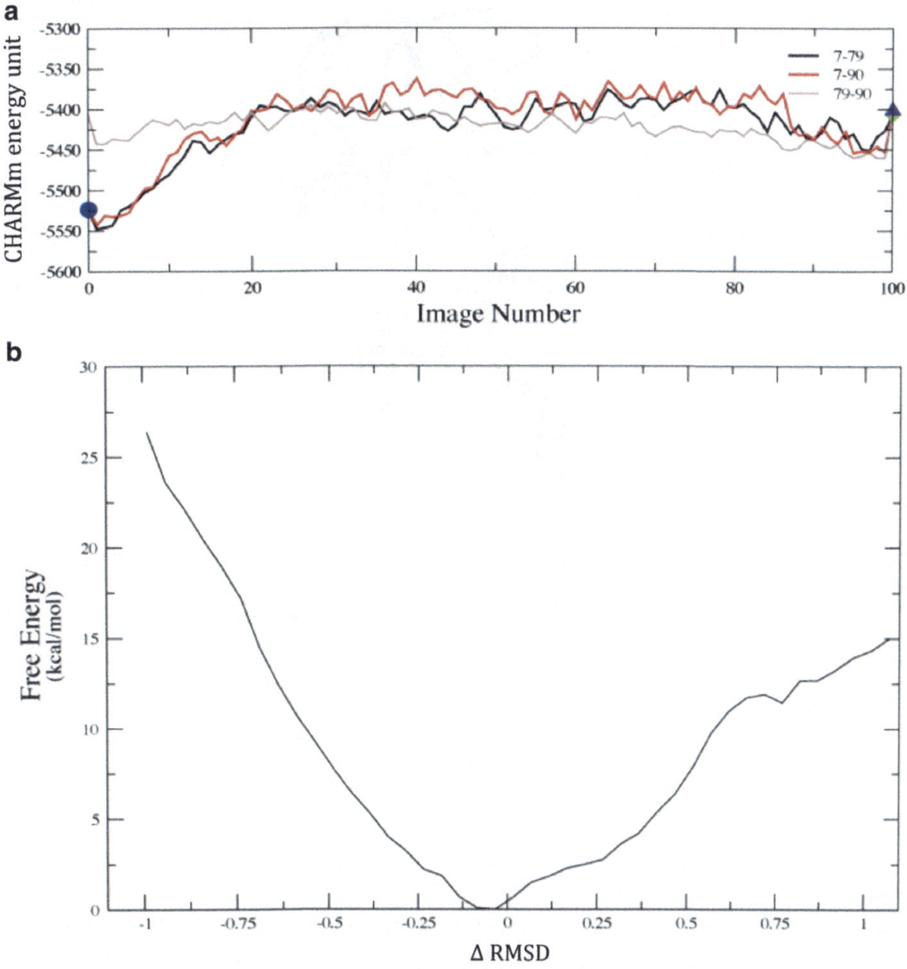

Fig. 6 (**a**) Converged average potential energies between stable, low-distance donor–acceptor conformations. The energies are computed using finite temperature string methods. Numerical labels are arbitrary names given to clusters of conformations that structural analysis showed to be identical. The x-axis, which denotes conformational variation, is the string image number. Reprinted with permission from J Chem Phys B 10.1021/jp207876k 2011. (**b**) The converged free energy corresponding to one of the potential energy curves above (*red curve* labeled 7–90). Note that the minima in average potential energy are on the edges of a rising free energy. The minima in (**a**) given in string images correspond to −0.50 and +0.69 Å in Δ-RMSD units along the string coordinate on (**b**). Reprinted with permission from J Chem Phys B 10.1021/jp207876k 2011

by fast sub-picosecond protein motions. More artfully put, one may describe the function of the enzyme as a *conformational filter* that performs a stochastic, but efficient search through conformation space that allows chemical reaction to happen in an almost barrierless fashion. The tradeoff in free energy is simply associated with the wait for the enzyme to locate a catalytically competent conformation in

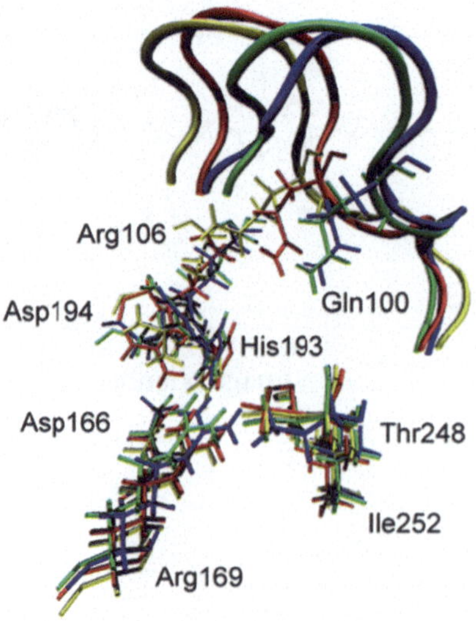

Fig. 7 Overlay of the active site residues of 4 representative potential energy stable conformations (three of which are part of Fig. 5a.) Residues belonging to the same conformation are shown with the same color. At the top of the picture, the active site loop is shown (residues 96–106, its residues 100 and 106 are shown explicitly). The 4 residues at the bottom of the figure are related to each other by a simple translation. However, the important residues His193, Arg106, Gln100, and Asp194 are being pushed by the active site loop (which has a different closed position in each of the 4 conformations), and their geometric arrangement is different in a non-trivial way among the 4 conformations. Transitions between these conformations were seen to always involve partial loop opening. Reprinted with permission from J Chem Phys B 10.1021/jp207876k 2011

which there is essentially no barrier to reaction. This focuses the view of enzymatic catalysis on the probabilistic rather than energetic component. It also highlights the value of mechanistic investigations such as have been described. It should be noted that this viewpoint is not a new "theory" or at variance with previous views of enzymatic catalysis [47, 48]. A completely equivalent way of stating this fact is that enzymes are catalysts and make chemical reactions go faster. The real scientific question is how do they accomplish this feat?

4 Rate Theories

None of the work we have described involves the computations of rates, but an important question is how our results might inform on methods of doing so. One of the most popular methods of rate calculation is transition state theory. Given the inherent complexity of a reaction in any condensed phase, especially that of an enzyme, it seems almost incredible that it can provide reasonable accuracy. In fact it often can, but involves treatments far more complex than the original theory. We now discuss both why it can provide useful rates and why modifications to the original theory are called for if it is to be applied to the types of reactions we have analyzed in this review. Finally we close this section with a description of other rigorous rate theories.

The original transition state theory or "Absolute Rate Theory" as proposed by Eyring [49] was based on the claim that there is an equilibrium established between the reactants and the transition state. Our TPS calculations clearly show that the transition state is far too ephemeral for such equilibrium to occur. This of course in no way invalidates the potential accuracy of TST, as it is well known that no equilibrium is needed to derive the approach [50, 51]. Truhlar has developed a comprehensive suite of methods that allow correction to the simplest transition state formulae [52]. There are two issues that at first glance would leave a skeptical critic thinking that TST had no chance of working. The first is simply that in a complex system the topology of the free energy surface is such that there should be very many saddles, and so it is likely that there should be very many transition states. In fact our calculations show convincingly that this is simply not what nature has arranged, at least as far as donor and acceptor pairs are concerned. In our calculations in LDH we have shown that although the reactive trajectories that are members of the transition path ensemble arrive at the transition state (the separatrix) in many ways, when there they are highly similar in characteristics, and the separatrix seems to be fairly low dimensional [53]. The concern may be raised that the search we performed for TPS in the LDH case was simply not broad enough. We have reinitiated these calculations with a far larger perturbation in the "shooting" move. We have never found members of the stochastic separatrix for either LDH or the "more floppy" DHFR that show wide conformational divergence. It should be emphasized that this was not a required outcome. There could have been multiple highly divergent paths through the separatrix. It should also be stated that it is always possible that the Monte Carlo sampling of TPS is incomplete in the cases we have studied, but given our attempts to find other regions of trajectory space have all failed, we suppose this not to be the case.

The second question one may pose is: if there is a promoting vibration, and a multidimensional reaction coordinate, how does one identify a transition state in low dimension? There is still a bottleneck; it is simply embedded in a higher dimensional space. The crucial point here is that the promoting vibrations we have found are motions on a similar timescale to that of barrier passage. They are not so fast that they can simply be averaged over to obtain a single free energy barrier, and they are not so slow that they can be assumed fixed for the purposes of rate calculation. In the language of transition state theory, these dynamics affect both the location of the TS and the height of the free energy barrier. We have discussed these issues and the relation to Grote-Hynes theory previously [54, 55]. In the generalized Langevin equation (GLE) formulation we presented in [52], the promoting vibration creates a GLE with a fluctuating reaction coordinate potential (i.e., the simple reaction coordinate has a free energy barrier modulated by another degree of freedom on a similar timescale to its own dynamics). The best analogy to these models in TST would be found in the ensemble averaged variational transition state methods Truhlar and co-workers have devised [56]. Barriers in which the promoting vibration had strongly compressed the donor–acceptor would have a lower probability in the ensemble of rates but have a greater contribution to the rate. In this view of TST, one cannot locate a single TS that accurately dominates the rate computation. It should be emphasized that this is a mathematical tool and not necessarily indicative of the actual stochastic separatrix.

One erroneous argument that has been made against the importance of dynamics in enzymatic chemistry is that since the TST recrossing statistic (κ) for enzymatic reactions has always been computed to be close to 1, there can be no dynamic effects. Dynamical effects such as promoting vibrations raise the rate of reaction. Their presence or absence could never be detected by the value of κ – as stated above, they are part of the free energy. For example, we have extended our barrier crossing calculations for many nanoseconds and have never seen a recrossing event – we are in total agreement that κ is 1.

Another question that has been raised is how much of an effect a fast motion such as the promoting vibration may have on the free energy barrier. Some years ago a rough calculation on HLADH suggested that the promoting vibration lowers the barrier by 30% over the static case [54, 55]. Clearly this will be significantly different from system to system. There is no clear way to test this number – there is not a simple mutation that breaks the promoting vibration, but we are investigating this possibility with experimental collaborators. The results described here in no way invalidate the possible overriding importance of such well-described effects as dipolar stabilization. On the other hand, we did try to initiate reactive trajectories in LDH with the promoting vibration restrained. This resulted in no reactive trajectories. This was not published because the restrained unsuccessful trajectories do not form a recognizable ensemble distribution.

Another point of contention that has been raised is how a motion so much faster than the rate of turnover could be involved in catalysis. This is in fact not surprising given that chemistry is almost never rate limiting in enzymatically-catalyzed reactions. Active sites are optimized to catalyze chemistry at a far greater rate than turnover, and it is likely that turnover is regulated for the biological system of interest. In a similar fashion, one would not be surprised that there is almost perfect charge complementarity in many active sites even though the chemistry this charge distribution catalyzes happens at a far faster rate than release of product from the enzyme. While enzymes never seem to have rates of turnover as fast as possible, they do seem to have individual steps that are highly optimized for their function. Whether that function is chemical change or substrate binding the release is at a physiologically relevant rate. In fact even when k_{cat} is measured, there are still protein rearrangements that dominate the measured rate. It must come as no surprise that passage of the chemical barrier is made possible by an optimized protein in much the same way that the optimized active site provides charge screening.

A final rate theory we would be remiss not to mention is the forward flux formulation of Chandler [57]. This is of course an exact formulation of the rate, much like the flux correlation formulation [58] is a formally exact method to compute the quantum rate. In these correlation function theories, TST-like results are recovered from a $t \rightarrow 0$ limit. The usefulness of the result depends on the choice of the dividing surface, as in TST. To obtain an accurate result in the enzymatic case, this dividing surface would have to be found for the full reaction coordinate, not simply particle transfer. So if there were a promoting vibration as part of the reaction coordinate, it would be included in the dividing surface search. It should be emphasized that such an approach does not guarantee accuracy – obviously the short time limit of the correlation functions involved is a drastic approximation to the rigorous full formulation.

5 Conclusions and Future Directions

It is time to see where this field stands 15 years after our suggestion that rapid motions of the protein matrix are in fact part of the reaction coordinate. From a computational perspective, the proposition seems well proven for both LDH and PNP on the basis of TPS computations followed by reaction coordinate identification. The continuing consternation in some quarters that the timescale of the promoting vibration is so much faster than the timescale of enzymatic turnover should not cause worry – in the cases we study the rate of enzymatic turnover is never limited by chemistry. In fact the rate of chemical barrier passage is very rapid, and on the same timescale as the promoting vibration. This perhaps begs the question of how the promoting vibration was "fashioned" by the evolutionary process because chemistry is not rate limiting, but the same is true for any aspect of chemistry in enzymes. Even though in the enzymes we observe today, chemistry is almost never rate limiting, and active sites have been carefully tuned to facilitate catalysis.

A continuing question is how these propositions can be tested experimentally. A method often used in mechanistic enzymology is mutation. While this is certainly a fairly blunt instrument for this question, it is a viable means of testing. As such we are currently performing TPS calculations on a number of mutants in which the promoting vibration has been disrupted. Interpretation of such results is always difficult, because mutation inside the protein seldom changes only one aspect of the structure; however it is to be hoped that such results will eventually inform on the involvement of promoting vibrations in the function of enzymes.

The Schramm group has developed a recent, highly promising, experimental technique [59]. Their idea was to create a "heavy" enzyme in which all non-exchangeable protein atoms are replaced by their heavy isotopic analogues. Termed a "Born-Oppenheimer" enzyme (because there is no change in potential energy, only vibrational structure) they have found changes of 30% in the rate of chemistry in the heavy enzyme. We are currently analyzing TPS calculations on the heavy enzyme to illuminate the specific changes in fast motion that causes this change in rates.

Finally, it is worthwhile to discuss future directions for this work. One critical issue is that of quantum dynamics in the reactive events. The TPS calculations we have done are completely classical in atomic dynamics. It has been suggested for many years that tunneling is critically important to a variety of hydride transfer enzymes, (see for example [60–62]) and so an important goal is the development of approaches that allow inclusion of tunneling in TPS simulations. Because TPS is an inherently classical phase space method, an approximate inclusion of quantum dynamics is needed. We have suggested employing centroid molecular dynamics as such a method [63], and are currently developing the capability to pursue such computations. The second crucial area we would hope to see addressed is the inclusion of protein dynamics in the protein design paradigm. The ab initio creation of artificial enzymes is obviously decades away, but limited enzyme engineering has been proven possible [64]. Since static design pictures, such as the catalytic antibody approach [65], have not been able to come close to matching natural enzyme's catalytic proficiency, it is worthy of speculation as to whether the subject of this volume – protein dynamics – could provide some of the missing functionality in such approaches. It is obviously difficult to envision exactly how these concepts could be made part of protein engineering design principles, but this may well be the next frontier for catalysis and protein dynamics.

References

1. Antoniou D, Schwartz SD (1997) Proc Natl Acad Sci USA 94:12360–12365
2. Austin RH, Beeson KW, Eisenstein L, Frauenfelder H, Gunsalus IC (1975) Biochemistry 14:5355
3. Agmon N, Hopfield J (1983) J Chem Phys 79:2042
4. Brooks B, Bruccoleri R, Olafson B, States D, Swaminathan S, Karplus M (1983) J Comput Chem 4:187–217
5. Gao J, Amara P, Alhambra C, Field M (1998) J Phys Chem A 102:4714
6. Riccardi D, Scheafer P, Yang Y, Yu H, Ghosh N, Prat-Resina X, Konig P, Li G, Xu D, Guo H, Elstner M, Cui Q (2006) J Phys Chem B 110:6458–6469
7. Cui Q, Karplus M (2002) J Phys Chem B 106:7927–7947
8. Hay S, Johanissen LO, Sutcliffe MJ, Scrutton NS (2010) Biophys J 98:121–128
9. Pudney CR, Hay S, Levy C, Pang J, Sutcliffe MJ, Leys D, Scrutton NS (2009) J Am Chem Soc 141:17072–17073
10. Johanissen LO, Scrutton NS, Sutcliffe MJ (2008) J R Soc Interface, Suppl 3:S225–S232
11. Johannissen LO, Scruttonn NS, Sutcliffe MJ (2011) Angew Chem 50:2129–2132
12. Dellago C, Bolhuis P, Csajka F, Chandler D (1998) J Chem Phys 108:1964
13. Dellago C, Bolhuis P, Chandler D (1999) J Chem Phys 108:9236
14. Bolhuis P, Chandler D, Dellago C, Geissler P (2002) Annu Rev Phys Chem 53:291–318
15. Dellago C, Chandler D (2003) In: Nielaba P, Mareschal M, Ciccotti G (eds) Bridging the time scales: molecular simulations for the next decade (vol 605 of Lecture Notes in Physics). Springer, New York
16. Hagan M, Dinner A, Chandler D, Chakraborty A (2003) Proc Natl Acad Sci USA 100:13922–13927
17. Vreede J, Juraszek J, Bolhuis PG (2010) Proc Natl Acad Sci USA 107:2397–2402
18. Onsager L (1938) Phys Rev 54:554–557

19. Antonio D, Schwartz SD (2009) J Chem Phys 130:151103
20. Antoniou D, Schwartz SD (2011) J Phys Chem B115:2465–2469
21. Ma A, Dinner AR (2005) J Phys Chem B 109:6769–6779
22. Peters B, Trout BL (2006) J Chem Phys 125:054108
23. Peters B, Beckham G, Trout B (2007) J Chem Phys 127:034109
24. Best R, Hummer G (2005) Proc Natl Acad Sci USA 102:6732
25. Schölkopf B, Smola A (2002) Learning with kernels: support vector machines, regularization, optimization and beyond. MIT, Cambridge
26. Basner JE, Schwartz SD (2004) J Phys Chem B 108:444–451
27. Basner JE, Schwartz SD (2005) J Am Chem Soc 127:13822–13831
28. Davarifar A, Antoniou D, Schwartz SD (2011) The promoting vibration in human lactate dehydrogenase is a preferred thermal channel. J Phys Chem B115:15439–15444.
29. Sagnella DE, Straub JE (1999) Biophys J 77:70–84
30. Sagnella DE, Straub JE, Jackson TA, Lim M, Anfinrud PA (1999) Proc Natl Acad Sci USA 96:14324–14329
31. Zhang Y, Fujisaki H, Straub JE (2007) J Phys Chem B 111:3243–3250
32. Lian T, Locke B, Kholodenko Y, Hochstrasser RM (1994) J Phys Chem 98:11648
33. Ghanem M, Li L, Wing C, Schramm VL (2008) Biochemistry 47:2559–2564
34. Saen-Oon S, Ghanem M, Schramm VL, Schwartz SD (2008) Biophys J 94:4078–4088
35. Saen-Oon S, Schramm VL, Schwartz SD (2008) Z Phys Chem 222:1359–1374
36. Saen-Oon S, Quaytman-Machleder S, Schramm VL, Schwartz SD (2008) Proc Natl Aacd Sci USA 105:16543–16548
37. Agarwal PK, Billeter SR, Rajagopalan PT, Benkovic SJ, Hammes-Schiffer S (2002) Proc Natl Acad Sci U S A 99:2794–2799
38. Dametto M, Antoniou D, Schwartz SD (2012) Barrier crossing in dihydrofolate reductase (DHFR). Mol Phys 110:531–536 (Bill Miller Festschrift)
39. Thorpe IF, Brooks CL 3rd (2005) J Am Chem Soc 127:12997–13006
40. Nashine VC, Hammes-Schiffer S, Benkovic SJ (2010) Curr Opin Chem Biol 14:644–651 and references therein
41. Becker O, Karplus M (1997) J Chem Phys 106:1495–1517
42. Pineda JRET, Schwartz SD (2006) Philos Trans R Soc 361:1433–1438
43. Pineda JRET, Antoniou D, Schwartz SD (2010) J Phys Chem B114:15985–15990
44. Ren W, Vanden-Eijnden E, Maragakis P, E W (2005) J Chem Phys 123:134109
45. E W, Ren W, Vanden-Eijnden E (2007) J Chem Phys 126:164103
46. Arora K, Brooks CL III (2007) Proc Natl Acad Sci USA 104:18496–18501
47. Nashine VC, Hammes-Schiffer S, Benkovic SJ (2010) Curr Opin Chem Biol 14:644–651
48. Hammes-Schiffer S, Benkovic SJ (2006) Annu Rev Biochem 75:519–541
49. Eyring H (1935) J Chem Phys 3:107
50. Pechukas P (1981) Annu Rev Phys Chem 32:159–177
51. Hynes JT (1996) In: Tapia O, Bertran J (eds) Solvent effects and chemical reactions. Kluwar Academic Publishers, Dordrecht, pp 231–258
52. Dybala-Defratyka A, Paneth P, Truhlar DG (2009) In: Allemann RK, Scrutton NS (eds) Quantum tunneling in enzyme catalyzed reactions. Royal Society of Chemistry, Cambridge, and many references therein
53. Quaytman S, Schwartz SD (2007) Proc Natl Acad Sci USA 104:12253–12258
54. Caratzoulas S, Schwartz SD (2001) J Chem Phys 114:2910–2918
55. Antoniou D, Schwartz SD (2001) J Phys Chem B105:5553–5558
56. Alhambra C, Corchado J, Sanchez ML, Garcia-Viloca M, Gao J, Truhlar DG (2001) J Phys Chem B 105:11326
57. Chandler D (1978) J Chem Phys 68:2959–2970
58. Miller WH, Schwartz SD, Tromp JW (1983) J Chem Phys 79:3759–3764
59. Silva RG, Murkin AS, Schramm VL (2011) Proc Natl Acad Sci USA 108:18661–18665
60. Bahnson BJ, Park DH, Kim K, Plapp BV, Klinman JP (1993) Biochemistry 32:5503–5507

61. Bahnson BJ, Colby TD, Chin JK, Goldstein BM, Klinman JP (1997) Proc Natl Acad Sci USA 94:12797–12802
62. Knapp MJ, Klinman JP (2002) Eur J Biochem 269:3113–3121
63. Antoniou D, Schwartz SD (2009) J Chem Phys 131:224111
64. Siegel JB, Zanghellini A, Lovick HM, Kiss G, Lambert AR, Clair JLS, Gallaher JL, Hilvert D, Gelb MH, Stoddard BL, Houk KN, Michael FE, Baker D (2010) Science 329:309–313
65. Stewart JD, Benkovic SJ (1993) Chem Soc Rev 22:213–219

Index

The manufacturer's authorised representative in the EU is Springer Nature Customer Service Centre GmbH, Europaplatz 3, 69115 Heidelberg, Germany. If you have any concerns regarding our products, please contact ProductSafety@springernature.com

Printed and bound by CPI Group (UK) Ltd, Croydon, CR0 4YY
30/09/2026
02225602-0003